W. H Cole

Notes on Permanent-Way Material, Platelaying and Points and Crossings

W. H Cole

Notes on Permanent-Way Material, Platelaying and Points and Crossings

ISBN/EAN: 9783337256401

Printed in Europe, USA, Canada, Australia, Japan

Cover: Foto ©berggeist007 / pixelio.de

More available books at **www.hansebooks.com**

NOTES

ON

PERMANENT-WAY MATERIAL,

PLATELAYING,

AND

POINTS AND CROSSINGS.

BY

W. H. COLE,
EXECUTIVE ENGINEER, P.W.D., INDIA.

"MENTE MANUQUE."

E. & F. N. SPON, 125, STRAND, LONDON.
NEW YORK: 12, CORTLANDT STREET.
1890.

PREFACE.

The permanent-way inspector of to-day—so far as my knowledge of him goes—is not merely that vague product, a "practical man;" he does not necessarily prefer guesswork to certainty; he has *not* a supreme contempt for a simple and intelligible mathematical method; he has *not* a serene confidence in the fallible process of "putting it in by the eye"; and he is not one jot less practical than the platelayer of the old school, although he has had the advantage of a better education. To him, then, I dedicate this little book.

The first chapter deals with the material which the platelayer has to work with. While on furlough in 1885 I was engaged for some months in passing material for the War Office, and so had constant opportunity of observing the processes of manufacture which I have briefly described. The inspection and test reports, I may add, are quotations from my own papers.

In the second chapter I have described the actual operations of platelaying, both as regards construction and maintenance; and for assistance in this portion I am greatly indebted to Inspector Campion and several other friends, both Engineers and Platelayers.

The third chapter treats of points and crossings. Upon this subject Donaldson is, without question, the great authority. But simplifications of his treatment follow at once from the measurement of the gauge from inside edge to inside edge of rail-head, instead of from centre to centre of rail, since the width of the upper table of the rail disappears from all terms of the calculation, and only affects the clearance. Donaldson, again, expresses the inclination of a crossing in terms of the base and height of an isosceles triangle whose equal sides are drawn on the centre-lines of the intersecting rails; whereas I have taken the inclination-number as expressed by the cotangent of the angle of the crossing. This is simpler, and not only disposes of the half-angle in most of the formulæ, but agrees with the practice of the Government of India, P.W.D.

The fourth chapter contains a few simple rules for setting out curves and diversions,—determining the radius of a curve and the super-elevation of the outer rail on curves,—finding the radius, lead, and number of a crossing, &c.

A few tables follow. Those which specially apply to the old and new standard crossings in use on Indian State Railways were worked out by me when I was Assistant Superintendent Way and Works, North-Western Railway, and, prefaced by a few simple rules, were circulated among all engineers and inspectors on that line. I wish to say, however, that my object has been, not so much to put ready-made figures at the disposal of the reader, as to indicate the simplest way of working out for himself any problem which may occur.

I regret that I cannot even attempt to give a list of the books which I have consulted, and from which, as I readily acknowledge, I have derived a great deal, either absolutely or in confirmation of my own experience. Engineers in India can only carry about with them a few standard books of reference; there are no libraries of professional works within easy reach; and all that we can do is to lend and borrow such few books as we have.

This little book has been written at scattered intervals of leisure, but I can only plead this as a partial excuse for its unequal treatment of the subject and lack of completeness.

W. H. C.

MULTAN, *November* 1889.

CONTENTS.

CHAPTER I.

PERMANENT-WAY MATERIAL.

ART.		PAGE
1.	Definition of Permanent Way	1
2.	Ballast	1
3.	Iron and Iron Ores	2
4.	Descriptions of Iron	2
5.	Cast Iron	3
6.	Malleable or Wrought Iron	4
7.	Steel or Ingot Iron	4
8.	Impurities in Iron	5
9.	Preparation of Iron Ores	5
10.	Reduction in the Blast Furnace	5
11.	Castings	6
12.	The Puddling Process	7
13.	Manufacture of Various Sections of Wrought Iron	8
14.	Manufacture of Steel.—The Cementation Process	8
15.	The Bessemer Process	9
16.	The "Basic" or Dephosphorization Process	10
17.	The Siemens-Martin Process	11
18.	Steel for Rails and Sleepers	11
19.	Manufacture of Steel Rails	12
20.	Inspection of Steel Rails	13
21.	Testing Rails	13
22.	Forms of Rails	15
23.	Manufacture of Steel Sleepers	16
24.	Testing Steel Sleepers	17
25.	Manufacture of Fish-plates	17
26.	Testing Fish-plates	18
27.	Chairs	18
28.	Fastenings	19
29.	Wood Sleepers	19
30.	Points and Crossings	20

CONTENTS.

ART.		PAGE
31.	Steel *versus* Wrought and Cast Iron as a Permanent Way Material	22
32.	Double-headed, Bull-headed, and Flat-footed Rails compared	22
33.	Longitudinal and Transverse Wooden Sleepers compared	23
34.	Metal and Wooden Sleepers compared	24

CHAPTER II.

PLATELAYING.

35.	Distribution of Workmen in Platelaying	27
36.	Number of Workmen	27
37.	Material Gangs	30
38.	Linking-in Gangs	30
39.	Lifting and Packing Gangs	33
40.	Duties of a Permanent Way Inspector	34
41.	Duties of a Sub-inspector	35
42.	Maintenance Gangs	35
43.	Duties of a Mate or Ganger	36
44.	Duties of the Wrenchman or Keyman	36
45.	Platelayers' Tools	37
46.	Maintenance of Rails	39
47.	Fishing the Rails, Expansion	40
48.	"Creep" or "Travel" of Rails	42
49.	Maintenance of Wood Sleepers	44
50.	Lifting, Packing, and Boxing	47
51.	Treatment of Metal Sleepers	48
52.	Curves	50
53.	Laying Points and Crossings	53
54.	Level Crossings	56
55.	Final Remarks	56

CHAPTER III.

POINTS AND CROSSINGS.

56.	Object of Points and Crossings	58
57.	Definitions	58
58.	Symbols	59
59.	Two Methods of Determining the Lead	59
60.	First Method. Lead of Crossing	61
61.	First Method. Radius of Crossing	61
62.	Second Method. Radius of Crossing	62

ART.	PAGE
63. Curve-lead	62
64. Switch-lead	63
65. Second Method. Lead of Crossing	63
66. The Platelayer's Rule for the Lead	63
67. Curves of Contrary Flexure	64
68. Curves of Similar Flexure	66
69. Lead of Crossing Constant	66
70. To Determine the Crossing	67
71. Modification of the Lead in certain cases	68
72. Turnouts and Crossovers	69
73. Gathering Lines	70
74. Three-throw Points and Crossings	71
75. Triangles	72
76. Crossing more than one Line; Diamond Crossings	73
77. Tables	74

CHAPTER IV.

SIMPLE RULES.

78. To set out a Curve by Offsets	75
79. To set out a Diversion from a Straight Line	76
80. To find the Radius of a Curve	78
81. To find the Cant required on a Curve	78
82. To find how much a Rail should be bent to suit a Curve	79
83. To find the number of a Crossing	80
84. To find the Radius of a Crossing	80
85. To find the Lead of a Crossing	80
86. To find the Crossing required in any case	81
87. To find the distance from Nose to Nose of Crossings in a Crossover	82

TABLES, ETC.

Report on Steel Rails	83
Tensile Tests of Steel Rails and Sleepers	84
Standard Types of Permanent Way—Indian State Railways	85
Table giving Cant for different Curves and Speeds	86
Tables of Points and Crossings, Turnouts and Crossovers, Integer Numbers	87
Tables of Points and Crossings, Turnouts and Crossovers—Indian State Railways	89

ERRATUM.

On page 30, line 9, *for* "strengthen" *read* "straighten."

ERRATUM.

Page 63, formula (7),

for—
$$L' = G \sqrt{(1+I^2)} + GI$$
read—
$$L' = G \sqrt{\overline{(1+I^2)} + GI}$$

NOTES

ON

PERMANENT-WAY MATERIAL,

PLATELAYING,

AND

POINTS AND CROSSINGS.

CHAPTER I.

PERMANENT-WAY MATERIAL.

1.—Definition of Permanent Way.

THE permanent way consists of rails, sleepers, and fastenings, and the term is used to distinguish the finished and permanent railroad from temporary lines laid down during construction for the carriage of material, &c.

The rails are laid to gauge, i. e. at a certain distance apart; their ends are fished together with plates and bolts; and they rest—either in chairs, on bearing-plates, or directly—upon the sleepers, to which they are fastened by means of spikes, fang-bolts, coach-screws, or keys.

The sleepers are bedded and packed in ballast, which is spread upon the formation of the bank or cutting.

2.—Ballast.

The ballast, forming the immediate foundation and support of the permanent way, should be of a material clean, hard, and sufficiently bindable, yet loose enough to admit of free drainage.

Stone, brick, or kunker, broken sufficiently small to pass in any direction through a ring 2½ or 3 inches in diameter—sharp clean gravel, or sand, is the most usual material.

Sand, covered with a top dressing of brick or stone to keep down the dust, forms an excellent packing for metal sleepers.

That portion of the ballast which is not actually packed, but lies on the top or is loosely filled in, is called the boxing. Continental engineers attach great importance to the boxing outside the rails forming heavy banquettes (*vide* frontispiece).

3.—Iron and Iron Ores.

In one form or another iron constitutes the material of rails, fish-plates, chairs, and fastenings,—and sometimes of sleepers and keys.

In a chemically pure state iron possesses no practical value, and is merely interesting from a scientific point of view.

Its qualities as a constructive material depend upon the more or less complete elimination of sulphur, phosphorus, silicon, and other impurities, and upon the admixture of very small proportions of carbon and manganese.

Iron is obtained in its first condition—that of pig iron—by smelting certain ores in a blast furnace with coke and lime.

Of workable ores, the oxide class includes magnetic iron ore, red hæmatite and brown hæmatite, while the carbonate class are found in the forms of spathic iron ore and clay ironstone.

Swedish iron is obtained from magnetic ore. Red hæmatite is found in Lancashire and Cumberland; brown hæmatite in the Forest of Dean, Cornwall, and Devonshire; and clay ironstone in Wales, Scotland, Yorkshire, Staffordshire, Derbyshire, and Shropshire.

4.—Descriptions of Iron.

Three descriptions of iron are broadly distinguished:—Pig or cast iron; malleable or wrought iron; and steel or ingot iron. Of each of these, again, several varieties are recognised.

5.—Cast Iron.

Cast iron contains from 2 to 4·75 per cent. of carbon, which in the whiter varieties is chiefly combined or dissolved, and in the greyer kinds mostly uncombined or graphitic. When the proportion is not too large, carbon increases the fluidity of molten iron, and makes the pig softer and tougher; in excess it makes iron both softer and weaker, owing to the presence of uncombined or graphitic carbon.

Cast iron is inductile, and cannot be forged or welded. It cannot be relied on to stand severe strains suddenly applied, and is untrustworthy under tensile strain.

Within a limited range it is tougher, in compression it is much stronger, and under tension it is much weaker, than wrought iron; it is harder, more brittle, and beyond a small range of stress it is less tough; it is fusible at much lower temperatures; and it is not so readily oxidized.

We may broadly distinguish three descriptions of pig iron—grey, mottled, and white—there being again four gradations of grey pig, of which three are useful for foundry purposes.

More particularly, then, we may say that there are six varieties of pig iron in the market:—

No. 1.—Contains most graphitic carbon, is the most fusible, is soft and tender, has the coarsest texture, with a fracture grey, crystalline, and lustrous, is used for small fine castings, and is generally the dearest.

Nos. 2 *and* 3.—Contain less carbon, are fusible enough for larger castings, where strength rather than fineness of surface is required, and exhibit a lighter grey and less lustrous fracture.

No. 4.—" Bright grey pig " contains least carbon of all the four grey kinds of pig, is not readily fusible, and is used, not for foundry purposes, but for making malleable iron. Its fracture is light grey with very small crystals and but little lustre.

Mottled pig.—So called on account of the appearance of its fracture.

White.—Contains most combined carbon, is the hardest and most brittle, exhibits a white, crystalline fracture, and is only used for making inferior bar iron.

6.—Malleable or Wrought Iron.

Malleable iron contains less carbon than cast iron, but not necessarily less than the milder kinds of steel; it is therefore best described as the result of the more or less complete decarburization and purification of pig iron by the "puddling process."

It is malleable (as its distinguishing name implies), ductile, weldable, and tough; it fuses with difficulty and only at a very high temperature; and rusts rapidly when exposed to the influence of a moist atmosphere containing carbonic acid. Its strength under tension and under compression is very nearly equal. The characteristic texture is fibrous, and this will generally be exhibited by fracture. The appearance may, however, be crystalline if the piece be broken by a sudden blow.

"Puddled bar" is the first kind produced in the manufacture. By repetitions of the process of cutting, piling, reheating, and rolling are obtained successively the descriptions of wrought iron known commercially as "merchant bar," "best," "double-best," and "treble-best"; and, as it is thus gradually improved in quality, it is made suitable for rails, ship-plates, boiler-plate, forgings, &c.

7.—Steel or Ingot Iron.

Steel, or more precisely "ingot iron," differs from malleable iron in that it is produced in a state of fusion, and from cast iron in that the ingot formed by the cast metal is malleable.

This statement affords perhaps a better differentiation of ingot iron from both cast and malleable iron than any which involves reference to its composition, or attempts to describe more particularly its special qualities. For the proportion of carbon, which it contains in a combined or dissolved state, may vary from $0 \cdot 1$ to $1 \cdot 25$ per cent., while in its character it also varies greatly, the milder kinds of steel resembling malleable iron in many respects.

The characteristic fracture of steel is bright and crystalline; but, when ruptured gradually, a mild steel may exhibit a fibrous fracture.

The milder kinds of steel, containing least carbon, resemble wrought iron in toughness, tensile strength, and weldability. The harder kinds are particularly capable of being hardened and tempered by heating and rapid cooling.

Steel is malleable and forgeable, but should be wrought at a lower temperature and treated more carefully than malleable iron. It is more fusible, but steel castings are rough, porous, and generally require to be forged.

8.—Impurities in Iron.

"Cold shortness"—or brittleness when cold—is commonly due to the presence of silicon, phosphorus, or arsenic; "red shortness"—or brittleness at a red heat—to sulphur, calcium, copper, and other impurities.

Phosphorus greatly reduces the tensile strength; but its hardening effect, and the increased fluidity which it induces in the molten metal are of advantage in light castings where great strength is not required.

Silicon, like carbon, affects the ductility of the metal, and makes pig iron weak, hard, and brittle.

9.—Preparation of Iron Ores.

The ore, if it contains much pyrites or shale, is broken up and weathered for three or four months, or even for a much longer period. It is then roasted in a kiln in order to expel from it, as much as possible, water, sulphur, and arsenic, and to render the ore porous.

10.—Reduction in the Blast Furnace.

The modern furnace is a huge vertical wrought-iron cylinder lined with brickwork. From the throat to rather below the middle, the stack or body of the chamber gradually bellies out. From here it contracts more rapidly in the boshes down to the hearth. At the bottom of the hearth is the tap-hole, which, while the furnace is working, is plugged with clay; the sides of the hearth are pierced to receive from three to six tuyeres, through which the hot blast is supplied from the main.

Once lit, the fires of the furnace may not be extinguished for years, charges of coke, lime, and ore being added from time to time through the throat.

In its gradual descent the ore is heated, reduced, and decomposed; water and carbonic acid are given off; carbonic oxide, hydrogen, and hydrocarbon gases deoxidize the ore; and the time, which acts as a flux, combines with the silica of the ore and forms slag.

The reduced iron now takes up a certain amount of carbon and some impurities as well; the metal fuses; the furnace having been tapped, the lighter slag flows off at a higher level; and the impure iron is either run into "pigs," whence it is known as pig iron, or the molten metal is carried at once in travelling ladles to the Bessemer converters to be made into steel.

As the temperature is higher or lower, and the quantity of fuel greater or less, the pig iron produced varies between the greyer and the whiter descriptions.

11.—Castings.

No. 1 is only used for small castings in which fineness, rather than strength is desired. For larger castings the greater hardness, strength, and toughness required, are obtained with No. 2 or 3, or a mixture of these; indeed, mixtures run more solid, and give better results.

The pigs are broken up and melted in a cupola furnace, charges of coke, iron, and limestone, being added as required, and a blast supplied by fans or blowers. When the furnace is tapped, the molten metal is caught in a ladle, poured into sand moulds, and allowed to cool gradually. The contraction in cooling is about one per cent. in linear dimension, and allowance must be made in moulding accordingly.

In a specification of cast-iron sleepers, it is not usual to prescribe what kind, or mixture of pig iron, shall be used, but to lay down certain tests. Thus, for Indian State Railway sleepers a cast-iron bar 2 inches deep, and 1 inch broad, placed on bearings 3 feet apart must bear, without breaking, a weight of 1½ tons suspended at the centre, i. e. the modulus of rupture must not be less than 3·375 tons.

Special castings, such as railway crossings, are "chill-cast" in moulds which are wholly or partly metallic, and by 'this means a hard skin is given to surfaces exposed to much wear.

12.—The "Puddling Process."

Malleable iron is produced by melting and "puddling" pig iron in a reverberatory furnace, the hearth of which is covered with broken slag, and the whole of the interior lined or "fettled" with a mortar composed of ground red hæmatite or some cheaper substitute.

White pig iron is preferred, and the charge (about 4 or 4½ cwt.) having been introduced, more coal is added, the furnace closed, and the heat raised.

In the "melting-down" stage the pigs, as the metal softens, are turned over with a long iron rod; silicon is partly separated, and grey pig is converted into white—graphitic carbon changing to the condition of combined or dissolved carbon.

In the next stage, during which the damper is lowered, and the temperature allowed to fall, the puddler mixes the charge with basic slags which are thrown in.

During the third stage the heat is again raised; carbonic oxide is freely given off in spirts of blue flame, and the metal is said to "boil." As the puddler works the mass vigorously with his rabble, most of the carbon, silicon, phosphorus, and manganese are exposed to oxidation and thus separated.

The "boiling" now subsides; the sinking mass becomes pasty; and granules of infusible iron begin to "come to nature." These, during the "balling stage," are worked into balls by the puddler, and drawn towards the fire-bridge, the cinder falling to the bottom of the hearth.

13.—Manufacture of various Sections of Wrought Iron.

The balls, having been taken out of the furnace, are welded under the shingling hammers, by which the slag is pressed out, and the mass consolidated.

While still hot, these "blooms" are passed through the "puddling" or "roughing" rolls, and the iron is then known as No. 1 or "puddled bar."

By repetitions of the process of cutting, stacking, reheating, and rolling the particular qualities of malleable iron required for rails, ship and boiler plates, forgings, &c., are successively obtained.

The proper section of plates, angles, tees, rails, &c., is finally acquired in the "finishing rolls."

For rails the stacks or piles are made up of puddled bars covered above and below with slabs of hard hammered iron, which after rolling, encloses the heads of the rails, and thus forms a good wearing surface.

Not only, however, are steel rails far superior to iron rails, but actually cheaper, so that steel has practically superseded iron in this as well as many other manufactures.

14.—Manufacture of Steel.—The Cementation Process.

"Blister steel" is produced by embedding wrought-iron bars in carbon and raising them, while protected from the influence of air, to a long-continued white heat. They thus become impregnated with a certain amount of carbon, but more particularly on the surface.

To obtain greater uniformity these bars are sheared, packed in faggots, raised to welding heat, and hammered or rolled again into bars, which are known as "shear steel"; and "double shear steel" is the result of repeating this treatment.

A more truly homogeneous steel is produced by melting blister steel in crucibles with the addition of manganese or spiegeleisen.

Shear and crucible steel are principally used for making cutting tools. It remains to describe more in detail the Bessemer, Basic, and Siemens-Martin processes, by which steel is now manufactured, possessing almost any desired gradation of qualities, at a comparatively cheap cost and in enormous quantities.

15.—The Bessemer Process.

Bessemer steel is obtained by the partial decarburization of pig iron—a grey pig iron being preferred, reduced from hæmatite ore and practically free from phosphorus or sulphur. The molten metal is carried in travelling ladles from the blast furnace to the Bessemer converters.

These are huge wrought-iron retorts, swung on trunnions, and having a capacity of from 8 to 10 tons. The bottom is removable, and fitted with perforated cylinders of fire-clay, through which the blast is introduced. The interior is lined with ganister, or with fire-brick laid in ganister mortar.

A wood fire is first lit inside the converter, and blasted out when the lining has become red-hot.

The converter is then swung down to receive its charge of molten pig iron, again raised, and the blast turned on. In about twenty minutes the charge is reduced to a state of almost pure iron.

Once more the converter is swung down, in order to receive the charge of molten "spiegeleisen," or ferro-manganese, which flows from a cupola close by, and is required to give the proper admixture of carbon and manganese.

After a few minutes the slag is partially run off, and the molten steel poured into a ladle. In the bottom of the ladle is a hole, whence the fluid metal flows (when the plug is lifted) into the ingot-moulds, most of the slag and other floating impurities being left behind.

The steel having solidified, the moulds are lifted off and the ingots removed to be re-heated and passed through the "cogging rolls."

16.—The "Basic" or Dephosphorization Process.

The acid or silicious slag produced in the Bessemer process has no affinity for phosphorus, so that this impurity cannot be eliminated. Now steel containing as much as 0·2 per cent. of phosphorus with a very low percentage of carbon would be utterly unfit for rails. It was not until Messrs. Gilchrist and Thomas introduced the "basic" process, in which the ganister lining is replaced by one of a strongly basic character, that it was possible to dephosphorize ores which could not be utilized in the Bessemer process.

The basic lining is usually prepared with dolomite, which is either burnt, ground, and applied with tar, or made into bricks and laid with tar.

Before introducing the molten pig iron and slag about 15 or 20 per cent. of burnt lime is thrown in, and with this a basic slag is formed which eliminates the phosphorus.

The blow lasts somewhat longer than in the Bessemer process; and it is during the "after-blow," when a very high temperature has been reached, that the oxidation of the phosphorus takes place, the carbon having been previously separated.

The ferro-manganese or speigeleisen is added and the operation continued as in the Bessemer process.

A very ductile steel may thus be obtained, but it does not generally exhibit so great a tensile strength as Bessemer steel.

17.—The Siemens-Martin Process.

The Siemens open-hearth steel-melting furnace is employed in this process. Below the hearth are the regenerators—two chambers filled with a chequer-work of firebrick.

A gas rich in hydrocarbons is generated from bituminous coal and other fuel in a separate apparatus and conveyed by mains and culverts to one of the regenerators. Here it absorbs the waste heat and, passing through, reaches the hearth of the furnace, where it mingles with the air which is necessary for its combustion, and which has been similarly heated during its passage through the other regenerator.

The charge consists generally of grey hæmatite pig, which is first thrown on the hearth, and of four or five times as much wrought iron or steel scrap, old rails, &c. The scrap melts in the more easily fused pig-iron, while the oxidizing flames play upon the whole mass all the while. After three or four hours samples are taken out and tested from time to time, and, when a satisfactory test-result has been obtained, about 1 per cent. of spiegeleisen is introduced and thoroughly mixed with the molten mass.

The fluid metal is then tapped, caught in a travelling ladle, and cast into moulds.

The marked advantages of this process are economy of fuel and the utilization of steel scrap.

18.—Steel for Rails and Sleepers.

The varying qualities of steel depend to a great extent upon the proportions of hardening ingredients—carbon, silicon, and manganese.

While manganese tends to lessen the red-shortness due to phosphorus, the alloy of manganese and iron renders the steel more liable to rust. The proportion of manganese should be lower therefore—its presence being less desirable—when the percentage of phosphorus and silicon is insignificant.

For rails a moderately hard steel is preferred,—containing not less than 0·3, nor more than 0·45 per cent. of carbon; not more than 0·06 per cent. of either silicon, phosphorus, or sulphur; and, besides these, no other ingredient except manganese and iron.

For sleepers, a milder steel, containing less carbon and manganese, is required. The proportion of carbon should not exceed 0·14 per cent., and sleeper-steel should be able to bear a tensional strain of between 26 and 30 tons per square inch, with a contraction of 30 per cent. of the original area at the point of fracture.

The following analyses of steel for rails, sleepers, and ship-plates may be given for purposes of general comparison; but it must be observed that test-results are more significant of the suitability of a steel for a particular use than its composition as determined by chemical analysis, for there may be considerable differences in the composition of steels equally adapted for certain requirements:—

Steel for	Rails.	Sleepers.	Ship-plates.
Iron	98·47	99·18	99·27
Carbon	0·35	0·11	0·14
Manganese	1·01	0·63	0·47
Silicon	0·05	0·02	0·02
Phosphorus	0·06	0·06	0·06
Sulphur	0·06	..	0·04
	100·00	100·00	100·00

19.—Manufacture of Steel Rails.

The ingot, having been reheated, is first passed through the "cogging" or "blooming rolls"; the ends sheared off; and it is divided, if necessary, either by shearing or under the steam hammer, into "blooms." The blooms, heated to redness, are then passed through the "roughing rolls," and lastly through the "finishing rolls."

While still hot, the long rail-piece is carried along live rollers and cut into the proper lengths by a circular saw.

When cool, the rails are straightened; the lengths checked, and, if necessary, corrected; and the holes are drilled.

The rails are then stacked for inspection.

20.—Inspection of Steel Rails.

The inspector is assisted by two or more men, who apply the gauge to every rail in each tier in order to check the lengths, and turn the rails over with nippers, so that he may examine them on every side and see whether there are any splits, cracks, cinder, or other defects.

Some must be rejected altogether; but those which are more than one-eighth of an inch too short, or exhibit flaws near the ends, may be marked to be cut.

When all the rails in one tier have been overhauled, the inspector looks along each one and marks those which require to be straightened. He should also apply templates, in order to test the section of the rail and the size and pitch of the fish-bolt holes.

While that tier of rails is being removed the inspector proceeds to examine the top tier of another stack.

Lastly, the inspector may have a few rails weighed, in order to compare the actual with the specified weight per yard.

21.—Testing Rails.

The composition and qualities of rail-steel are proved by:—
Chemical analysis; direct tensile test; the dead load test; and the falling weight test.

The results of a chemical analysis of a rail-steel, as compared with that of steel for sleepers and ship plates, have been given in Article 18.

At the end of the book are given examples of an inspection report, with details of the dead load and falling weight tests applied to certain of the rails inspected. For the purpose of comparison, moreover, the results of tensile tests applied to pieces of rails and sleepers are shown in another table.

The direct tensile test is applied in a testing machine of either the hydraulic or lever type. A test-piece of a certain diameter, having been turned out of the head of a rail, is subjected to a gradually increasing tensile strain until it is torn asunder. The breaking strain, the difference between the original and fractured areas of section, and the elongation or difference in length between two marks on the bar before and after rupture are then recorded. The tensile strength is thus measured, and the ductility of the sample is indicated by the contraction and elongation. A rail-steel should be able to bear a tensile strain of from 30 to 33 tons per square inch.

The dead load or lever test is also applied in the testing machine. A piece of rail, resting on bearings a certain distance apart, is subjected to a gradually increasing load or pressure, and the temporary deflection is measured from time to time until a permanent set takes place. The deflection indicates the elasticity of the metal until the limit of elasticity is reached, after which the test may be continued until the breaking weight finally determines the transverse strength of the rail, while the deflection which the rail will bear up to rupture shows the ductility of the metal.

The falling weight test consists in letting fall a certain weight from a certain height upon a whole rail or crop-end supported on bearings a certain distance apart. For the Indian State Railway flat-footed 62 lb. steel rail, the weight, fall, and bearing were 1 ton, 15 feet, and 3 feet respectively; and the rail was to bear two such blows without showing the least sign of fracture. For the metre gauge flat-footed $41\frac{1}{4}$ lb. steel rail, the weight of the monkey, the fall, and the bearing are

½ ton, 15 feet, and 3 feet. The deflection after each blow is measured and recorded.

Of all these tests the last is considered by Messrs. Sanberg & Snelus to be the most significant and important. It tries the metal suddenly and severely, as it is tried in actual use, and tells at once whether the steel is too soft or too hard.

The results of a tensile test applied to a piece cut from the head of a rail, and to a piece turned out of the web or lower flange would probably show that the tensile strength differs over different parts of the section. Moreover, as Mr. B. Baker pointed out, the tensile and dead load tests in themselves are not such as prove the capability of the rail to stand the strain of loads applied not only instantaneously, but, so to speak, by shocks and blows.

22.—Forms of Rails.

There are three forms of rails in common use—the "Vignoles" or flat-footed, the double-headed, and the bull-headed.

These are all modifications of the I section, a form suggested by the work which a rail has to do, considered simply as a girder.

The edges of the upper table must, however, be rounded to suit the wheel-tires; allowance must be made for wearing down; and the mass of the head is distributed rather in depth than in width. In the lower flange of "Vignoles" rails the width and flatness are retained.

The fishing angle should be as small as possible; in other words, the slopes of the upper and lower shoulders should be as flat as they can be rolled conveniently, in order to increase the direct support of the fish-plates and diminish the strain on the bolts. Sometimes the width across the shoulders is greater than that across the top of the rail, with intention to prevent slackening of gauge as the rail gets worn and to lessen the action on the wheel-tires on curves.

The proportions of the rail ought to be such as to allow the rolled rail to cool evenly. Finally, regard not merely to wear, but to the jars and shocks to which a rail is subjected by heavy trains oscillating and pounding over it, demands a thicker web and a far heavier section throughout than would be necessary from our primary consideration of the rail as a simple girder. The proportions of weight in the head, web, and flange of Mr. Sandberg's "New Goliath Rail" are 45·5, 22, and 32·5 per cent. respectively. The head is not so deep as in the older pattern, but is wider. The interior wears so much more rapidly than the surface metal, that a deep head is less economical than a wide one which offers a larger surface of contact between the tire and the rail. At the same time it gives that better bearing for the fish-plate which, in the American type, was obtained by sloping out the sides of the head to make the shoulders broader than the upper table. The American form is also intended to prevent slackening of gauge as the rail gets worn, but it certainly increases the friction on the wheel-flanges as well.

If the rail is to be reversible, the lower flange must be exactly the same as the upper, and we thus obtain the double-headed rail, which does not rest directly upon wood sleepers, but requires the intermediate support of cast-iron chairs.

Again, the lower table, resting on the seat of the chair, frequently becomes so dented that it is scarcely serviceable when reversed. The form of the double-headed rail may be more or less retained with, however, a lighter lower flange, and thus the bull-headed section is developed. This enables us to retain also the chair, which affords the more efficient lateral support required to stand the shocks of heavy and rapid traffic.

23.—Manufacture of Steel Sleepers.

The steel sleeper, recently adopted on Indian State Railways is trough-shaped, the ends as well as the sides being turned down to hold the packing. The top is strengthened by a thicker longitudinal strip, $4\frac{1}{2}$ inches wide, running along the middle, from end to end. Two lugs or clips are stamped out of the top of the sleeper on each side of the rail-seat; one of these grips the flange of the rail on one side, while on the other side a steel key is driven between the rail and the clip, so that the foot of the rail is firmly held by the clips and key. Slack gauge on curves may be given by driving one or both of the keys on the inside instead of the outside.

The plates are reheated, shaped in a powerful press, and the clips stamped out. At the same time a tilt of 1 in 20 is given to the rail-seat. It is important to press all the sleepers at very nearly the same heat, in order that there may be no sensible difference of gauge. When cold the sleepers are dipped in a bath of tar. Instead of being rolled first in flat plates, the steel may at once acquire the trough section in the mills, and then be pressed in the die as in the former case.

There must be no burrs on the clips—or the key will not drive fairly—and the plates must be free from corrugations.

24.—Testing Steel Sleepers.

The sleepers having been laid out on benches, the inspector selects a few, has rails fitted on them and keyed up, and tests the gauge. He has a certain number flattened out under the steam hammer, when flaws in the shoulders may be detected; and the clips are tested by being doubled back. Should there be more than 0·14 per cent. of carbon in the steel, the clips will fail under this test. Samples are also cut out of selected sleepers and subjected to tensile strain in the testing machine; examples of this test are given at the end of the book.

25.—Manufacture of Fish-plates.

The fish-plates are rolled in the mills, cut to the required lengths, punched hot or drilled cold, and straightened in a press. The ends are filed to remove burrs left by the saw or shears, and the plates, having been dipped in drying oil, are secured together in bundles of twelve.

Rails, especially if made of wrought iron, tend to split or spread at the ends, which wear down as the (so to speak) unsupported particles on the edge slip away. To afford a stiffer joint and prevent this sinking and flattening of the ends of the rails, therefore, such forms as the angle or the double-deep fish-plate are sometimes adopted. The ordinary fish-plate has only about one-third the strength of the rail section.

In order to hold the heads of the bolts, while the wrench-man is screwing them up, the fish-plates are indented so that the bolts cannot slip; or the neck of the bolt is made square or oval, to fit holes of the same shape in one of the fish-plates.

26.—Testing Fish-plates.

The section of the fish-plate and the size and pitch of the bolt-holes are tested by templates; test pieces are subjected to tensile strain in the testing machine; and selected plates are bent to a right or more acute angle on the anvil, in order to prove that they will bear this treatment without tearing.

27.—Chairs.

Cast-iron bowl or pot sleepers are so designed as to form a combined sleeper and chair for double-headed or bull-headed rails, but with wood sleepers the intermediate support of a cast-iron chair is required. The rail generally rests on a seat, and fits tightly against one jaw when the key, which is usually of pressed oak, is driven between the rail and the other jaw.

It is found that the lower tables of double-headed rails are often so chair-marked that it is useless to reverse them; and accordingly in one form of chair the head is suspended in the jaws, the lower table of the rail not touching the chair. In this case, however, the blow of a passing train is not transmitted to the rigid chair through the more elastic medium of the whole section, but only through the head, and this loss of cushion is a disadvantage affecting both the rail and chair. A better remedy is a broader chair seat.

Pressed wooden keys cannot be expected to last more than five years, especially in a country where the hygrometric changes are so rapid as in India. Spiral metal keys are sometimes used, but it may still be said that no efficient substitute for the pressed wood key has yet been discovered.

28.—Fastenings.

The rail or chair is fastened down to the wood sleeper by spikes, coach-screws, or fang-bolts and clips.

Chairs may be fixed to the sleeper with round spikes.

Fang-bolts are liable to work loose. The thread rusts; and in screwing up, the bolt may break, or, if the wood be soft, the fang loses its grip, and the fang-bolt is merely turned round uselessly in its hole. Coach-screws are better, but they also rust in the thread and work loose.

The rectangular dog-spike, with bearing-plates either throughout or at the joint-sleepers only (except on bridges and curves), makes an excellent fastening for the flat-footed rail on wooden sleepers.

29.—Wood Sleepers.

Hard wood is preferred. Sleepers should consist of thoroughly sound and seasoned wood, as free as possible from sapwood, large or loose knots, shakes, cracks, or defects of any kind.

Fir and pine are in Europe the most abundant and common material, oak being far superior to either, but correspondingly expensive. In India, deodar is generally used for the ordinary road, and sâl or teak for bridge and crossing sleepers; creosoted pine is imported, however, from Europe, and may be actually cheaper than deodar within a certain distance from the port of landing.

There are several processes by which railway sleepers may be more or less preserved. The efficiency of the "kyanizing process" is doubtful, as the corrosive sublimate is liable to be washed out by excessive moisture. "Burnett's process" is open to the same objection when the chloride of zinc solution is weak enough not to affect the strength of the wood. In France "Boucherie's (sulphate of copper) process" has been largely used; while in England sleeper wood is generally creosoted. Ordinary sleepers of deodar and other indigenous woods last fairly well in India, if the climate be a dry one, without any protection at all; but it is usual to cover bridge-sleepers of teak or sâl with tar.

The sleepers are generally arranged transversely. In section they may either be half-round or rectangular, the latter as a rule. If the rail be bull- or double-headed, the inward tilt of the rail is given by the peculiar form of the chair, and that portion of the sleeper on which the chair rests is merely truly planed; but where the flat-footed rail rests directly upon it, the sleeper is adzed to an inward slope of 1 in 20, so that the rail has a corresponding cant of 1 in 20 from the vertical inwards.

If the sleepers are arranged longitudinally, as is still the case on a few railways, they are connected by transoms and bolts at intervals.

30.—Points and Crossings.

The object of points and crossings is to pass trains from one line of metals to another.

The points or switches are short, tapering steel rails, so adjusted to their respective stock-rails that when one switch is pressed against its stock-rail the other is drawn away, and thus one line of metals or the other is made continuous.

At the intersection of the inner rails of the two tracks a crossing is placed, and the rails diverge in front of the nose of the crossing, so as to act as guard-rails and at the same time leave a gap for the train to cross on either line. Guard-rails are also fitted to the outer rails opposite to the crossing.

The heel of the switch either rests in a double heel-chair, or is fished to the adjoining rail and stock-rail (with a wedge between), in such a way as to allow the switch to pivot with sufficient freedom on the heel. The switches are held rigidly together by two connecting rods, to one of which the point-rod is attached; this leads to a bell-crank connected with a counterweighted lever by turning over which the switches are moved one way or the other on the slide-chairs; and as the switches are not fastened to the latter but merely rest on them, they require the lateral support of studs which are passed through the stock-rails at intervals and, pressing against the web of the switches, prevent the latter from bending in when a train is going over them.

Switches or tongue-rails are made of ordinary steel rails, selected and machined.

The following is the complement for a pair of switch and stock rails, flat-footed, metre gauge :—

    2 tongue-rails.
    2 stock-rails.
    12 slide-chairs, with stud-bolts and cotters.
    2 connecting rods.
    1 lever-rod or point-rod.
    1 counterweighted lever and crank.
    2 stock-wedges.
    8 bolts for ditto.

For double-headed switch and stock-rails, 5' 6" gauge, the following is the complement of parts :—

    2 tongue-rails.
    2 stock-rails.
    2 toe chairs.
    6 slide chairs, No. 1.
    2   „   „   2.
    2 heel chairs, No. 3.
    2   „   „   4.
    6 slide chair bolts and nuts or cotters.
    1 throw-over lever.
    2 connecting rods.
    1 hook rod with bolt and pins.

The points should be fitted with a locking-bolt, lock, and key.

Crossings are frequently made of cast steel in one solid piece with the wings, and so formed that the upper and lower sides are similar and the crossing is reversible. Built-up crossings, however, of ordinary steel rails, selected, machined, and fitted together are much better and much cheaper.

31.—Steel versus Wrought and Cast Iron as a Permanent Way Material.

The superiority of steel to iron as a constructive material is based upon the higher limit of elasticity, greater ultimate tensile strength, and greater ductility which it possesses. On the other hand, it corrodes more rapidly under adverse atmospheric conditions, and most rigid care and supervision are required in its manufacture to prevent brittleness and ensure toughness.

We must look for other qualities, however, in the case of rails; and here the supersession of iron by steel is principally due to the homogeneity and ability to withstand heavy moving loads without being crushed, abraded, or laminated, which distinguish the latter material. The life of a steel rail is very probably twice as long as that of an iron rail.

Steel castings are much more expensive than iron. They require to be annealed; the moulds are more costly, the melting-point being so much higher; there are more dead-heads; the castings are rougher; they are more liable to contain blow-holes; and they require to be forged. For cast sleepers and chairs, therefore, iron is preferred. The comparative merits of rolled and pressed steel sleepers, of cast-iron sleepers, and of wooden sleepers, will be dealt with further on.

32.—Double-headed, Bull-headed, and Flat-footed Rails compared.

The strong lateral support afforded by chairs leads at once to the adoption either of double-headed or bull-headed rails wherever the permanent way has to carry heavy loads moving at great speed.

In India, however, and on the Continent, where the speed and oscillation of trains are not so great, the flat-footed rail held down by dog-spikes, with or without bearing-plates, is generally equal to all requirements. Held by lugs and keys in one or other of the forms of pressed steel sleepers recently adopted, the flat-footed rail is as securely fastened as one can desire.

On most English railways the bull-headed rail in chairs has found favour. The reversibility of the double-headed rail is held to be of doubtful advantage, because the lower table is liable to corrode in brackish soil, and is frequently so dented by the chair that the rail, when inverted, is at its best only fit for sidings. Moreover, the tensional strain to which it has been subjected for years as the lower table, makes that part of the rail less able to bear the compressive strain which it suffers as an upper table when the rail has been inverted, and it fails rapidly.

The first objection has been met with but indifferent success by suspending the head of the rail in the jaws of the chairs, for in this case the shock of heavy trains is concentrated on the head of the rail only, instead of being distributed through the whole mass of the rail. The only effectual prevention of chair-marking is good keying, firm packing, and a clean chair-seat. If these are ensured, the economy of a double-headed rail, with four edges to wear out, seems to be clearly proved.

In flat-footed rails with comparatively large heads—the thin foot and the compact head receive such different amounts of work in the rolls, and cool down when rolled at such different rates, that the interior of the head is liable to prove spongy—and the work of straightening the rail requires very special care.

33.—Longitudinal and Transverse Wooden Sleepers compared.

The longitudinal sleeper affords a continuous bearing, so that a lighter rail may be adopted. On a broad gauge road this arrangement may actually require less timber than the transverse. The first cost, however, will generally be greater because the scantling is greater; repairs are less easily made; and the rail lies with the grain and cuts into the wood.

The transverse arrangement is at once more convenient, more economical, and almost universally adopted.

On abutments of bridges a longitudinal sleeper-frame may be introduced with advantage to carry trains easily and smoothly on to the girders.

34.—Metal and Wooden Sleepers compared.

Not only do wrought iron and steel corrode more rapidly than cast iron, but the result of this corrosion is more rapidly destructive to sleepers of those materials, the metal being much thinner.

Wrought iron as a material for sleepers may be considered as discarded in favour of steel.

The steel sleeper is still on its trial. The form adopted on Indian State Railways has already been described. Howard's pattern has not long since been introduced on the Great Northern Railway of England with the 82 lb. bull-headed rail. It is trough- or rather channel-shaped, being open at the ends. Two tongues or lugs are cut out and pressed down, affording an elastic seat for the rail, which is secured by a wooden or metal key. We have, therefore, steel sleepers adapted for the flat-footed and others for the bull-headed (or double-headed) rail, and in each of these there are only three parts—the sleeper proper, and two keys. The fastening is at once simple and efficient.

Whether steel sleepers will stand the test of time in brackish soil remains to be seen. The fact is, the rusting of metal sleepers generally has been, on the whole, exaggerated, and such preventives as galvanizing, steam oxidation, and oxide of lead are seldom required. Sleepers kept in reserve, or exposed to such adverse atmospheric conditions as are met with during carriage by sea, should, however, be tarred or painted.

Cast-iron pots have been known to last 25 or 30 years in India on a sandy soil, and in a very dry climate. In other parts, where the soil is clay, and the rainfall considerable, a pot road is difficult to maintain.

The wrought-iron cross-ties, gibs, and cotters, which hold the cast-iron pots together, rust more rapidly, and are taken out and renewed with difficulty.

Old cast-iron pots can be melted down and recast either into sleepers or chairs, while steel and wrought-iron scrap are practically useless in India.

Some patterns are adapted to a key driven inside, with the idea that loose keys are then more easily detected by the inspector. This arrangement is obviously less convenient for the keyman, who is also engaged in screwing up loose fish-bolts on the outside of the rail. Moreover, the inside key is slackened by a passing train, whereas the outside key is tightened and forms an elastic cushion.

The greater weight of metal sleepers, as compared with wood sleepers, is so far an advantage that it gives a solid and steady foundation, and is, on the other hand, a drawback, when the matter of long carriage has to be considered.

Wood sleepers will no doubt hold their own where timber is plentiful and cheap. Yet it is significant that, in countries so rich in timber as Germany, Sweden, and Norway, the use of metal sleepers is increasing, chiefly, of course, to husband the forest resources. Timber affords an elastic support, and in station yards, where derailments are frequent, the damage caused by such accidents is far less than with metal sleepers. Laid in good clean kunker or stone ballast, deodar sleepers will last 12 or 15 years.

On the whole, comparing metal with wood sleepers, it may be said that the average life of the former is twice as long; that their first cost will not always be much greater; that even when unserviceable, they have a certain value as scrap; that with metal sleepers, the gauge is better kept; that their maintenance is cheaper; and finally that, after a year or two, their renewals are more regular, whereas those of wood sleepers increase year by year, and are generally most irregular.

By way of illustration of the comparative first cost of four different classes of permanent way at Karachi (on the sea-board), and at Lahore (827 miles from the sea-port Karachi, but within easy reach by river of the hill-forests), the following figures may be of interest:—

	Karachi. Rs.	Lahore. Rs.
75 lb. F.F. steel rails, deodar sleepers, joint bearing plates, and spikes	19,150	19,850
75 lb. F.F. steel rails, pressed steel sleepers, and steel keys	20,180	22,880
73 lb. B.H. steel rails, cast-iron pot sleepers, with ties, gibs, cotters, and pressed wood keys	20,450	23,850
73 lb. B.H. steel rails, deodar sleepers, chairs, coach-screws, and pressed wood keys	21,720	22,730

In regard to fastenings for metal sleepers, Sir Guilford Molesworth is in favour of vertical wedges of cast iron. Wrought-iron wedges rust and get jammed. Screws rust in the thread. Rivets work loose and rattle.

CHAPTER II.

PLATELAYING.

35.—Distribution of Workmen in Platelaying.

THE platelaying of a new line of railway cannot be commenced until the banks and cuttings have been made up to the proper formation, the bridges completed (or temporary expedients arranged in their place), and permanent way material collected at suitable depots.

The operations connected with the platelaying may be divided into three :—

(*a*) Conveyance of material to the working-point;
(*b*) Linking-in ; and
(*c*) Lifting, straightening, packing, and boxing.

Accordingly, the supervisional and working staff must be regularly organized in three divisions :—

(*a*) Material gangs ;
(*b*) Linking-in gangs ; and
(*c*) Lifting and packing gangs.

36.—Number of Workmen.

Whatever the number of men employed may be—and that must vary with circumstances and the speed with which the work has to be pushed on—it is essential that the men be thoroughly well organized and supervised ; every man should have certain work to do; and there should be no overcrowding or confusion.

Supposing that the material is brought up and laid out rapidly and continuously, the speed of platelaying is only limited finally by the time required to link two pairs of rails together, and the numbers and distribution of workmen depend upon this entirely.

While it might take three minutes to link-in a pair of 21-feet rails, a gang of ten men could not lift, pack, fill in, straighten, and box up the same length of road in less than fifteen minutes; and the faster the linking can be done the more men will be required to lift and pack within the same time. Although no exact proportions, therefore, can be fixed, it may be said generally that the material gangs and the lifting and packing gangs ought to be at least three or four times as strong as the linking gang, if all the operations are to be carried on at an even rate.

In platelaying the Kandahar Railway—perhaps the smartest piece of work of the kind on record—the rails were linked together, when the work was thoroughly in hand, at the rate of two pairs per minute, whereas at first it took two minutes to link a pair. The average progress over $133\frac{1}{2}$ miles was $1\frac{3}{4}$ mile per diem.

Unless time is the chief consideration, however, great care and accuracy in laying a new track will be amply repaid by easier maintenance afterwards, and far longer life of material. The platelayer's chief object should be to lay the road in a workmanlike manner, rather than to rush it through. His rails should be straightened or curved as required, his sleepers laid square and properly spaced, the gauge exact, expansion carefully allowed for, his bolts cleaned and oiled, and every preliminary arrangement made to ensure good platelaying.

Referring to the actual platelaying only, the following, including bheesties, would probably be able to link-in and finish one-quarter of a mile of double-headed rails on cast-iron pot sleepers per diem, on the 5' 6" gauge:—

Fixing tie-bars	5
Linking-in	65
Filling sand for lifting and packing up	65
Lifting and packing	50
Levelling sand and putting on a 3 in. top dressing of ballast	65
	250

PLATELAYING, AND POINTS AND CROSSINGS.

For such a gang the following tools would be required:—

Bars, large	20
Bars, small	10
Baskets	150
Chisels (for splitting cotters)	5
Hand hammers	5
Keying hammers	5
Lifting levers	2
Phaoras	125
Pot rammers	100
Sling-hooks (and bamboos)	12
Spirit level	1
Straight edge with height boards	1
Wooden beaters	20
Wooden square	1
Wrenches	4

To link-in about half a mile per diem of metre gauge road, flat-footed $41\frac{1}{4}$ lb. steel rails on deodar sleepers, the number and distribution of the workmen (including trollymen, bheesties, and artificers) might be as follows:—

Mate in charge of trollies	1
Trollymen with material trollies	24
Rail-carriers	16
Sleeper-carriers	40
Mate in charge of carriers	1
Men cleaning and oiling fish-bolts	4
Boys carrying-fastenings, placing the spikes on the sleepers, &c.	4
Augermen, boring the sleepers and spiking	8
Men assisting augermen, holding up the sleeper to the rail with bars, &c.	8
Wrenchmen, fishing the rails together	2
Men assisting wrenchmen	4
Mate in charge of linking	1
Bheesties	2
Carpenters, making tool handles, nicking the sleepers, &c.	2
Blacksmith	1
Hammerman	1
	120

Three material trollies will be enough if the lead does not exceed half a mile. The first, manned by eight trollymen, will be loaded with sleepers; the second, loaded with rails, will require ten trollymen; and the third, carrying fastenings, six only.

The linking gang enumerated above will be followed by a packing gang consisting of a mate, an assistant mate, and sixteen men.

The mate will strengthen the road and lift the joint-sleepers; after which the assistant mate will level the intermediate sleepers, and see that the road is properly packed.

37.—Material Gangs.

These are engaged in unloading the material from trains and loading it on carts or trollies; in again unloading the material and distributing it alongside the line; and in clearing the line, as the working-point advances, of surplus material. The regular and continuous supply of material is, as already observed, of the first importance.

38.—Linking-in Gangs.

A portion of these will first of all be engaged in placing approximately in position the material which has already been delivered alongside by the material gangs, laying the sleepers and rails as nearly as possible in place, and putting the fastenings where they can be got at readily.

The fish-bolts ought to be opened out, cleaned, and oiled before they are laid out for the wrenchmen.

One line of rails should be linked and spiked first. The first party of wrenchmen half-fish the rails, adjusting for expansion; while the rest complete the fishing of the rail-joints.

These are followed by the augermen, who gauge the rails, bore the sleepers, and spike the rails or chairs; and are assisted by levermen, who by means of crowbars or rail lifters hold up the sleeper to the rail while it is being spiked.

The sleepers should be properly spaced by the mate, one rail being marked with chalk at the proper intervals, and these marks squared off on the other rail.

The road is then roughly straightened, lifted, and packed.

A clearer idea of the details of linking-in a new road will, perhaps, be obtained from the description of an actual day's work.

Let us suppose that a quarter of a mile of 24-ft. flat-footed iron rails on wood sleepers is to be taken out and replaced by 30-ft. flat-footed steel rails of heavier section on steel trough sleepers.

The new material has been carefully laid out on one side, so as to leave room for throwing out the old material on the other. Some of the rails had been sagged by careless handling or by being tipped up in short trucks, the ends of which could not be let down to allow the rails to lie flat. These have since been straightened in a portable rail-press, and all are now laid out in pairs, in two straight lines at the edge of the ballast, the ends not touching but having the proper clearance for expansion between them—one-quarter of an inch or perhaps less—so that the new rails occupy the exact length required right along the line. On the slope of the embankment, opposite the centre of each pair of rails, lie the proper complement of steel sleepers, two pairs of fish-plates, and eight fish-bolts; the boxes of steel keys have been placed in convenient positions; and all the tools have been arranged ready for use at the foot of the slope.

The inspector has not more than 100 men, although he could find work for at least 250. He has to let a train pass at 9.30 a.m. before breaking the road, and must finish the job before 6 p.m.

Meanwhile the men are engaged on the road which was renewed the day before. One mate and gang are rough-straightening; another mate and gang are finishing the straightening. One lifting gang is taking up the joints and middle;

another is setting the road, putting a finishing top on it, and boxing up. The wrenchmen are tightening up the fish-bolts.

Another gang is stripping the ballast from the road which is to be renewed to-day, removing all but two fish-bolts at each joint, chipping sleepers to rail-seat level so that the claw-bars may grip the heads of the spikes to be withdrawn, easing the spikes and splitting the fang-clips with a cold sett and heavy hammer. Boys are engaged in picking up the fastenings which have been taken out, oiling fish-bolts, chalk-marking the sleeper distances on the new rails, &c. At the end of the portion of road to be renewed are some 15 or 20 pieces of rail of different lengths, from which closers may be selected without having to cut and waste rails. The ends of these are already drilled to take the fish-bolts; and if the closers are of various lengths from 4 inches to 7 feet, say, with two pairs of slotted fish-plates, most gaps may be closed without cutting long rails or pulling back the old road beyond. When such short pieces are dropped in and fish-plates with long slots used, precautions should of course be taken to strengthen weak joints by placing pieces of old sleepers under them, until they are removed in continuing the renewals the next day.

After the last train has been cautiously passed over the partially dismantled road, the closers and specially made fish-plates (connecting the old and new rails) are removed and trollied to the further end, the remaining fish-bolts taken out, and the outer spikes drawn.

The old rails are then pinched out, lifted off with the rail-nippers or tongs, and deposited on one side.

The sleepers, which often stick on the old bed, are tipped up with light pinch-bars slipped under the spike-heads, or with picks the ends of which catch in the spike-holes. They may then be easily lifted by coolies and laid on the slope of the embankment, top side uppermost, so that the remaining spikes may be drawn at leisure. Of the spikes and fish-bolts perhaps 70 per cent. may be recovered as serviceable.

The new steel sleepers are then placed, and the right-hand

rails. The latter are first of all rested upon wooden blocks about 10 inches high, a pinch-bar being held between the ends, to keep the expansion-clearance approximately, while the rail is lined in, and one end of the sleepers lifted and clipped on to the rail by the lugs.

After the wooden blocks have been removed, and the rail has been dropped into place, a clearance-wedge is inserted between the ends, and the rail driven close up. One of the sleeper-keys at each joint must always be driven before the fish-plates are fitted on, as it cannot afterwards be driven from the fish-plate side.

The opposite rails are then levered on to the sleepers, pinched into the lugs, and hammered home against the inner lugs, while a man at the end of each sleeper levers it up to the rail.

The wrenchmen now half-fish the joints with two bolts (or even one only if there is much press for time) to hold the wedges in. They check the squareness of the joints with a set-square, and the clearance-wedges are allowed to remain in until the fish-plates have been completely bolted up for a distance of eight or ten rail-lengths ahead by the following wrenchmen.

The keys are then driven. Some of the lugs having perhaps got bent, it may be necessary to drive a chisel first and knock it out before the key can be fitted. The chisel has been so shaped that it will also serve as a drift for the fish-bolts. Impracticable keys are at once picked out to be straightened, or to have the burrs chipped off with a chisel, but, as a rule, it is found that the keys drive easily without using the drift-hammer.

It now only remains to put in the closers at the end of the relaid track, and to pack and straighten the road before passing the first train over it.

39.—Lifting and Packing Gangs.

These will perfect the operations of straightening, lifting, and packing; the first party straightening the road, and lifting and packing the joint sleepers; the second party levelling up the intermediate sleepers, and packing the road sufficiently to carry material trains over it at a speed of 15 or 20 miles an hour.

40.—Duties of a Permanent Way Inspector.

We now come to the matter of maintenance, and the permanent way inspector must remember that he is personally responsible for the state of the road.

It should be his object to make the road both safe and elastic; to effect this as economically as possible; and, only when he is sure that he has a really good road, to make it look like one.

He must personally superintend renewals, utilizing his material to its utmost; and either he or one of his sub-inspectors should always be present when a rail is being renewed.

The inspector has to keep an account of all material laid in the line, taken out and stacked, or received from stores and held in reserve for purposes of renewal.

He should regularly and minutely examine all bridges and culverts, points and crossings, signals, &c., so that he may be able to certify that they are in good order; and he should see that side drains and water-ways are cleared of long grass and rubbish.

The inspector ought to trolly over the whole of his section at least once in three days, carefully noting all bad rails and sleepers, loose or broken fastenings, and other defects. With this object, in boxing the ballast, spikes and other fastenings are to be left bare. He can best detect weak places and feel the quality of his road when travelling on the engine, and this he will frequently do, either going or returning.

When he has to take his trolly off to let a train pass, he should observe how the train affects the road, and will often thus detect weak joints and loose sleepers which would otherwise escape his notice.

He will of course stop at each gang to give orders, and should occasionally stay with them some time, direct their work, and see whether the mate understands his duties.

When arranging for any special work, he should have a sample piece done, time the men, and thus judge the time required to complete it.

His inspections should be made irregularly, so that the gangs may not know when to expect him.

He should report all accidents to the officials concerned, and bring to the notice of the assistant engineer any important or dangerous defect in the permanent way, bridges, &c.

The inspector should get thoroughly acquainted with his men, and know exactly what each man can do.

When a train is approaching, the gangmen should form up on one side of the line, so that the inspector or engineer, if he is in the train, may see that all are present.

41.—Duties of a Sub-Inspector.

It is the duty of a sub-inspector to carry out the inspector's orders implicitly. He will have charge of a sub-section of line, and must be present whenever a rail is being changed or heavy lifting being done, giving due notice and arranging danger signals.

He should always carry with him on the trolly a supply of fog-signals, two large red flags, hand-lamps, spanners, two chisels, a hammer, a ratchet-brace and bits, a gauge, and a telegram-book.

42.—Maintenance Gangs.

An average over the whole line of two men (including wrenchmen) on the metre gauge, or of three men on the 5' 6" gauge should be sufficient to maintain the permanent way when once it has been got into working order.

On the metre gauge a mate and gang could maintain at least three miles; while, on the 5' 6" gauge, more men being required together to handle the heavier material, a length of four or even five miles would not be too much, and would place twelve or fifteen men under the control of each mate.

The head-quarters of the gang should be as nearly central as possible, and close to the line.

43.—Duties of a Mate or Ganger.

The mate has direct charge of all the ordinary work of maintenance. He has to take out and replace all defective material, to stack old material at a sufficient distance from the line, to patrol the line in time of danger from rain or any other cause, to watch the action of water-courses in flood, to see that signal and telegraph wires are not obstructed in any way, to keep all water-channels open and clear them of rubbish, to repair the fencing, to clean the working parts of points and signals, to take charge of all tools and materials in his length, &c.

With a section of four or five miles the mate, if his head-quarters are fairly central, should generably be able to walk over his beat, half one day and half the other. He can scarcely do more than this, for the gang cannot be trusted to work by themselves, and his place is with them.

When his line is in fair condition he should make his men work—week and week about—from one end or the other.

44.—Duties of the Wrenchman or Keyman.

The keyman is the mate's right-hand man, and generally able to take up his duties in his absence. As a picked man he usually gets a little more pay than the rest of the gang.

Starting early in the morning, the keyman should go over the whole of his beat—driving loose keys and spikes, tightening loose bolts, and replacing broken fastenings. He should then begin again at a certain point, and do a quarter of a mile or so thoroughly.

The keyman on a pot or chair road carries the following tools and fastenings with him:—

    1 key-hammer, 6 lb. or 7 lb. in weight.
    1 large spanner.
    1 small spanner.
    6 fish-bolts and nuts.
    6 wooden keys.
    1 small hand chisel.
    1 steel scraper for cleaning chair-seats.

On a road of flat-footed rails on wooden sleepers, the wrenchman would carry the following:—

    1 hammer, about 4 lb. in weight.
    1 spanner.
    1 small hand chisel.
    6 fish-bolts and nuts.
    6 spikes.

It is as much as he can do to work through the whole beat properly, and he will seldom be able to return to work with the gang. Indeed, the more he is kept to his particular duties the better, instead of being encouraged to rush through his length and rejoin the others. If he goes through rapidly and carelessly, he will pass a good many loose keys, split others with a hurried stroke, and take little notice of loose bolts and broken fish-plates.

He should bring all defects of the road to the notice of the mate. If the latter is otherwise engaged, the keyman may take a couple of men through with him, if necessary, to pack loose sleepers.

45.—Platelayers' Tools.

The following lists give the full complement of tools required (1) for a gang consisting of one mate, one keyman, and eight men on a three-mile length, 5' 6" gauge, B.-H. rails, and cast iron pot sleepers; and (2) for a gang consisting of one mate, one keyman, and eleven men on a four-mile length, 5′ 6″ gauge, F.-F. rails, and wooden sleepers :—

Tools and Appliances.	B.H. Rails and Pot Sleepers.	F.F. Rails and Wooden Sleepers.
Augers, ½ in. or ⅝ in.	—	2
Baskets	8	11
Beaters	2	11
Book of railway regulations	1	1
Clawbars and crowbars	6	8
Cold chisel	1	1
Flag signals	2 sets	2 sets
Fog signals	12	12
Gauge	1	1
Hand hammer	1	—
Hand lamps	2	2
Jim Crow, or rail straightener	1	1
Keying hammers	2	—
Lifting lever	1	1
Munj rope	100 feet	100 feet
Phaoras	8	11
Ratchet brace and bits—	1	1
Screw wrench	1	1
Sighting boards	2	2
Sling hooks	2	2
Spanners	2	2
Spiking hammers	—	2
Spirit level	1	1
Steps, or inch boards	2	2
Straight edge	1	1
W.I. punners	8	—

The tools should be collected and counted by the mate every night, and kept in a locked tool-box.

It is scarcely necessary to say that gauges are apt to wear loose, and should be tested every two or three months by the inspector.

A short steel-pointed chisel-bar—or "tommy-bar"—is a useful tool, and very handy for turning over rails when loading, unloading, or shifting them. For lifting and carrying rails a sort of pincers, gripping the head of the rail and having horizontal handles, will be particularly useful in platelaying. With six such nippers, twelve men will pick up or lay down a 30-ft. 75-lb. rail both quickly and gently, and carry it with the greatest ease. While the unnecessary multiplication of tools is to be avoided, no tool need be regarded as superfluous which saves time and labour, and enables the workmen to handle material more easily and less roughly.

46.—Maintenance of Rails.

When one edge of the head of a rail is worn, the rail may be turned end for end in its place, or shifted to the other side.

A double-headed rail, the upper table of which is worn, may be inverted, unless the lower table has been corroded in brackish soil or dented by the chairs. It will at any rate serve to put in a siding, from which good rails may often be taken for the main line.

Double-headed rails frequently become chair-marked through neglect, the chair-seat not being kept free from rust and dirt. It is a good plan to make the mates open out their road—say two or three rail-lengths at a time—unkey and raise the rail, clean the seats thoroughly, replace the rail, and drive the keys, first those in the middle, next those at each end, and lastly the intermediate ones. Some of Greaves' oval pots were fitted with wooden packing-pieces to prevent chair-marking.

Turned rails do not wear anything like so well as the original upper table.

Fish-plates, from neglect, sometimes get a set camber, the wheels drop heavily on the low joints, and the ends of the rails are rapidly flattened or split, unless the fish-plates are inverted in time.

Rails which are worn or split at the ends may be cut, and

used as closers or check-rails. Every possible use ought to be made of a rail before it is finally rejected as unserviceable.

When a rail begins to fail, a stake is driven at the side of the line, and the deterioration of the rail carefully watched until it is necessary to remove it.

Rails should not be taken out and replaced during the heat of the day. If, from insufficient room for expansion, the inspector thinks that the rails will close in somewhat, when one is taken out, he should have the rails pulled back from either side the day before. Otherwise, he may have to use a cut rail. Inspectors have sometimes ten or twenty rails to change in one day at different places. Accordingly fish-bolts and nuts are unscrewed, cleaned, oiled, and replaced a day or two before the new rail is to be put in, so that the actual renewal may be easily and quickly done.

Either the inspector or sub-inspector must invariably be present on such occasions, and see that the line is protected by hand and fog-signals on each side.

Before breaking the road the inspector should make sure that the new rail is of the exact length required, with sufficient clearance for expansion.

47.—Fishing the Rails. Expansion.

When two rails are being fished together, a small space is left between them to allow for expansion. This space may be from one-eighth to one-fourth of an inch, according to the extreme variations of temperature, the season of the year, and the time of day when they are linked-in. Wooden or angle-iron wedges or iron rings of different thicknesses, must be placed between the ends of the rails while they are being fished, and are easily removed with a bar when the bolts are screwed up.

If too much space be allowed for expansion, the fish-plates and bolts will often snap in the cold season, and the bolts may be so jammed against one side of the holes that it will be difficult to take them out, if required. To lessen the space, the bolts should be slackened during the hottest time of the day, so that the rails may expand freely; the bolts may then be screwed up again.

On the other hand, if too little space be allowed, the rails will either buckle or creep. Should they buckle, the road may be temporarily adjusted to a reverse curve until there is an opportunity of opening out the road, taking out the fish-bolts, oiling them, and replacing them on one side of the joint only, loosening the chair-keys, drawing the rails sufficiently apart to admit the expansion wedges, removing one pair of rails, and replacing them by short rails as closers. The wedges should not be taken out until the rails have been eased for a considerable distance ahead.

Rails 30 feet in length are not so easily handled as shorter ones, and unless loaded in special rail trucks are more liable to get sagged, when carried by rail. But, while we are considering the subject of expansion, the further objection to rails of such a length is suggested by the fact that more clearance for expansion is required, the joints are, therefore, more likely to sink, and consequently the ends to be spoiled.

Fish-bolts should be cleaned and oiled before they are put in, and they should not be screwed up too tightly, not only lest they should break, but because the tight lateral grip of the fish-plate may prevent free expansion, and either cause the rails to creep or buckle.

Fish-bolts are frequently damaged by the use of too long a spanner, or by neglecting to oil them before screwing or unscrewing them.

When a new rail is put in between two old ones, the heads of which are worn down at the ends, it may be necessary to put in liners or packing-pieces to raise the latter.

On a curve the inner rails follow a necessarily shorter line, so that the inner rail-joints tend to get ahead of the outer joints. When this lead is as much as three inches, a rail cut six inches shorter may be put in on the inner side. In this way the inner joint will be thrown three inches behind the outer, and twice the distance will have to be covered before the lead of the inner joint is as much as three inches and another cut rail is required.

Rail-joints are to be avoided as much as possible in small open top or girder bridges, or where a check-rail is to be put in.

48.—" Creep " or " Travel " of Rails.

This is one of the greatest difficulties with which the engineer and platelayer in India have to contend, and a most troublesome problem to deal with.

The crux of the whole matter lies in the too tight lateral grip of the fish-plates, which prevents each rail from expanding independently, and causes a "creep" to be rigidly transmitted along the line of rail in the direction of least resistance. If the sleepers are firmly anchored and the fastenings hold the rail so that it cannot slide on them, the rails tend to buckle ; if not, they must creep. Sometimes this tight frictional grip follows from the particular form of the rail and fish-plate. The latter may be too long, or else, instead of only being in contact under the shoulder and upon the foot of the rail, it may fit closely to the web also, so that when screwed up it presses tightly against a large surface of the side of the rail. But the chief fault will often lie with the adoption of a section of rail having too great a fishing angle—in other words, the slopes of the upper and lower shoulders are not flat enough. In this case the fish-plate does not form such a direct support, and the bolts have to be tightly screwed up. The fishing-angle is frequently as great as 60°, but should not be more than 30°.

When, however, we come to inquire in which direction the creep takes place and why, it is almost impossible to give a satisfactory reply. On a double track the tendency of the creep will probably be in the direction of the traffic on each road ; on a single track in the direction of the heavier traffic, or from a higher to a lower level.

Moreover, this general movement or ploughing up of the rails by heavy traffic in one direction or the other will almost invariably affect the right and left lines of rails unequally, so that the joints are thrown out of square to a serious extent. Not only so, but the right and left-hand rails often travel in opposite directions.

In the case of a chair road the resistance of the chairs is of the nature of a couple, and the effort of expansion must tend to back and loosen the keys of one side so that the two rails are most free to travel in opposite directions. If the chairs were right and left handed, the keys could be driven in the same direction on both sides, and one rail would not necessarily travel faster than the other by backing out the keys and overcoming their resistance. If again the key-seat of the jaw were tapered from both sides to the middle of the jaw (as Mr. Mallet, Chief Engineer, P.W.D., proposed), instead of from one side to the other, the key would be gripped most tightly at the middle of the jaw, and could be driven from either side indifferently. Mr. Mallet pointed out that this would enable the joint sleepers to be spaced closer together, which would materially relieve the strain on the fish-plates. But there would be the further advantage that the key could be driven on each side in the same direction as the creep, and thus assist in resisting it. This method involves the use of two kinds of keys, one right and the other left-handed.

Steel sleepers are best able to resist the creeping movement, for their sharp edges grip the foundation of the ballast, and the rail is held very tightly by the lugs and keys. Here, therefore, the tendency is to buckle, and it is often difficult to keep a perfectly straight road, however heavily ballasted.

Wood sleepers move more easily, but, even if they hold, the rail may slide on the spikes, unless the fish-plates are notched to hold the heads of the spikes at the joint sleepers.

The rounded shape and smaller surface of frontage of pot sleepers afford the least resistance of all to the creep movement, but supposing them to stand their ground, the whole matter depends finally upon good keying.

To correct the unequal travel of the right and left lines of rails is at once a tedious and costly operation. The fish-bolts for some distance have to be taken out, oiled, and replaced, the rails to be pulled square in opposite directions, and one pair has to be taken out and replaced by two cut rails as closers.

Little more has been done, it must be acknowledged, in the preceding remarks than to set forth the difficulties and complications which beset the question of creep. Scarcely a single explanation can be put forward in any one case which is not apparently contradicted by the opposite circumstances of another.

49.—Maintenance of Wood Sleepers.

In laying a line, the joint sleepers are spiked first and then the intermediates. Before spiking down a chair, the key is driven tight. The sleeper is then placed evenly under the chair, square to the rail, and properly spaced, and is firmly held up to the chair with a crowbar while the hole is being bored and the chair spiked to the sleeper. When one side has been bored and spiked, the gauge should be put on (at right angles to the rail on the straight and normal to the rail on a curve), the sleeper squared and held up, and the opposite side bored and spiked to gauge. On a curve the outer rail is the one to be spiked first.

When the rail is simply held by spikes the sleepers must be adzed under the rail to an inward slope of 1 in 20 to tilt the rail and thus make it fit the coning of the wheel-tires. This is unnecessary when chairs or bearing-plates are used; these must not be counter-sunk in any way; the cant is given in the plates or chairs, and all that is required is to level and dress the portion of the sleeper on which they rest.

In boring a sleeper, an auger of rather smaller size than the required hole should be used for deodar and other soft woods, so that the spike when driven may fill the hole completely. If the point of the auger will not hold it must be oiled, not forced. All holes for spikes must be drilled right through the sleeper. The rectangular dog-spikes should be driven vertically and square to the flat-footed rail. Should the sleeper-wood be very hard and liable to split, the spike may be lubricated with soap before it is driven.

Bearing-plates and spikes make an excellent fastening for flat-footed rails, but, unless the traffic is fast and heavy, it will generally be sufficient to use the bearing-plates at joint sleepers only on a straight road, provided that the sleepers are of good quality.

When a sleeper is no longer fit for use at a joint, it may still do service as an intermediate. The joint sleepers are the most severely tried, owing to the joints, however good, yielding more than the body of the rail; the middle sleepers next require careful attention; the remainder are less important. It is in this order that the renewals of sleepers, therefore, should generally be carried out, rather than in continuous lengths; and economy can be secured by shifting partly worn sleepers from one position to another of less importance.

It is seldom feasible to turn over a sleeper in India, for the under side is not only torn by sharp-pointed beaters or by the ballast-packing, but often partially destroyed by white ants, even while the top is still fairly good.

If the bases of chairs do not cover a sufficient area there is a tendency to cut into the sleeper, the more so as the long edges of the chair run in the direction of the wood-fibres. If, however, the keys are kept tight and the ballast is sound and well packed, there will be less tendency to rock, and the chairs will not dent the sleepers so badly. Even when thus injured the sleepers will serve their time, if undisturbed. When a key is found to be slack, the chair-seat ought to be thoroughly well scraped and cleaned before the key is driven tight again.

All old spike-holes should be filled with tarred wooden plugs, neatly driven in. When the spike-holes are badly worn or rotten, the sleeper may be shifted a little transversely, so that new holes may be drilled. But if only one chair on a sleeper requires renewal, the old spike-hole should be plugged and re-bored for the new chair, the opposite chair not being touched.

When a new sleeper is put in, the old bed of ballast should be beaten up first, and the boxing should not be filled in for some days, during which the sleeper may require a little packing.

Sleepers which are getting into bad condition should not, as a rule, be covered over with ballast, but exposed until it is necessary to remove them. In a dry district, however, this may not be advisable, for fear of the sleeper being burnt by droppings from the fire-box of the engine.

Wood sleepers should be opened out for inspection after the rainy season, and will thus, moreover, be thoroughly dried.

When sleepers get out of square the spikes should be drawn, the sleepers replaced at right angles to the rail, the gauge applied, and the sleepers respiked.

If a loose sleeper is neglected for some time, the rail will begin to fail at that place; every train will drop heavily on the loose sleeper and aggravate the mischief; and the rail, which might have lasted for years, is rapidly spoiled. While working regularly through his length, therefore, the mate should also attend to these isolated cases of loose sleepers. The gang can pack a certain portion of road, say a furlong, in two days. On the third the mate or wrenchman should take one or two gangmen over the whole length to pack loose sleepers, while the rest box and dress the ballast.

Old sleepers, when taken out, should be piled in small stacks at some distance from the line and from each other; earth should be thrown on the top of each stack, and the ground close by cleared of grass or other rubbish liable to catch fire from sparks from a passing engine.

In laying bridge sleepers, the distance between girder and rail is to be carefully measured from point to point. The sleepers should then be cut accordingly on the lower side and marked for their places, so that differences due to camber of girder, rivet-heads, and cover-plates may be allowed for.

50.—Lifting, Packing, and Boxing.

The use of the height-board or straight-edge and sighting-boards should first be explained.

The height-board should not be placed further than 300 feet away from the lifting-point as a rule. The rails across which it is placed and those at the lifting-point should, first of all, be brought to exact level by means of the spirit-level.

The height of the sighting-boards (which are used to determine whether intermediate points are to be picked up or lowered) corresponds with the white line of the height-board or straight-edge. The latter (on the 5' 6" gauge) is a vertical plank, 6 feet long, about 8 inches deep, and fitted with feet at either end to steady it on the rails.

One sighting-board is placed by the lifter, and over this he sights the white line of the straight-edge with the intermediate sighting-board, which is placed on every successive joint, and indicates whether the line is to be raised or lowered.

The ballast consists generally of broken stone, brick, or kunker, if the sleepers are of wood; and of sand, covered with a top dressing of broken brick or stone, or wholly of the latter, if the sleepers are of steel or of cast iron.

Sleepers should be well packed directly under the rail-seat, and to a distance of 10 to 18 inches on each side according to the gauge; the middle and ends should be only loosely filled with ballast.

Gangmen, unless prevented, are much given to driving a big piece of stone or kunker under the sleeper, from which treatment the sleeper suffers considerably, while the apparent firmness of the packing is but temporary.

Since it is far easier to lift than to lower a road, less ballast is spread as a foundation on the formation than will ultimately be required under the sleepers.

The joint sleepers should first of all be packed, then the intermediates, and lastly the joint sleepers again before the road is boxed up.

In boxing, the rails and fastenings should be left bare; a light top dressing of ballast may be laid on wood sleepers to protect them from fire; and at least sufficient ballast must be spread beyond the ends of the sleepers to cover them freely when the ballast takes its natural slope.

A new or lifted road should not be fully boxed for some days, in order that it may settle and be packed up a little, if necessary.

In lifting a wood sleeper road, it should first be stripped of ballast to make it as light as possible, and thus prevent the spikes from drawing when the road is forced up with heavy levers. The rail, not the sleeper, should be held up with the lever, and both rails should be lifted together, the levers being held in place until the joints are firmly packed, the intermediate sleepers being then packed, and lastly the joints again beaten up. The old bed ought always to be broken up and renewed with fresh ballast. Four men should be put on to pack a sleeper, and they should beat the ballast under both sides equally.

A road must not be lifted more than three inches at a time, the elevation being made up, if necessary, by repeated lifts of not more than three inches. The lift should be made towards approaching trains on a double line, and from a lower to a higher level on a single track, the elevation being worked off at each end with an easy grade.

In picking up slack joints, the ordinary crowbar will be found to be handy and quite suited to the purpose.

51.—Treatment of Metal Sleepers.

The trough-shaped steel sleeper is packed in very much the same way as the wood sleeper, i. e. the ballast is packed tightly under the rail seat, and to a distance of 10 or 15 inches on each side, the middle being loosely filled, and at least sufficient ballast spread beyond the ends to cover them, and keep the ballast filling from slipping out under the edges.

The steel keys should be driven with a light hammer and the ends split.

In renewing steel sleepers the old bed ought always to be dug up and renewed with fresh ballast. Steel sleepers newly laid require considerable attention until they have become firmly bedded.

A cast-iron pot sleeper put into the road as a renewal is apt to break, if packed up too much, and to secure an even bearing it is advisable to open out and repack one or two sleepers on each side. Tie-bars should be gauged beforehand, and keys and cotters driven with a light hammer.

When pots "blow," the best way to secure the packing is to open out the earth to the rim of the pot, pack small ballast under it with picks, fill in earth or sand from the top as usual with punners, and with the addition of a little water this will hold firm.

A clay packing will probably set hard and shrink during the hot weather, and should be broken up, or the pot may split.

The importance of good keying is obvious. Pot sleepers at joints often get out of square and tighten the gauge. This is due to neglect on the part of the keyman.

Not unfrequently the key is so carelessly driven that the rail is canted too much inwards and the edge of the rail sheered off by the jaw.

The rail should be exactly fitted in its proper position, and the key, if firmly driven, then acts as a cushion as well as wedge.

Accumulation of rust and dust between the rail and chair-seat and slack keying are the chief causes of the under table of rails being dented. The breakage of pots may also be traced almost invariably to slack keying.

In lifting a pot road, all filling and ballast should be removed and replaced with fresh packing. When the rail is canted too much inwards from careless keying, the lifting should be done by applying the crowbar from the inner side of the rail. Indeed the rails ought to be unkeyed, lifted from the chair-seat while the latter is scraped and cleaned, and the key carefully redriven.

52.—Curves.

As previously explained, short rails have occasionally to be put in on the inner side of curves to correct the advance of the inner rail-joints.

To determine the exact amount of slack gauge to be allowed on sharp curves, we must find what is the versed sine on a chord whose length is that of the maximum rigid wheel-base plus the distance beyond to the points where the wheel-flanges touch the inner edge of the outer rail. To do this we may apply the formula

$$v = \frac{3 C^2}{2 R} \text{ (approximately)} \ldots (1)$$

where
  C = chord expressed in feet,
  R = radius of curve, also expressed in feet, and
  v = versed sine in inches.

Thus, on the 5 feet 6 inch gauge, with a maximum rigid wheel base of 16 feet, the allowance would be liberal, if we assumed C to be 20 feet, and this would give us—

$$v = \frac{600}{R}.$$

Accordingly, the allowance for slack gauge might be as follows:—

R	v
600 feet	1 inch
800 ,,	$\frac{3}{4}$,,
960 ,,	$\frac{5}{8}$,,
1200 ,,	$\frac{1}{2}$,,
1600 ,,	$\frac{3}{8}$,,

while curves of 2000 feet radius or more should be spiked or keyed to straight gauge.

Similarly on the metre gauge, with a maximum rigid wheel base of 10 feet, the formula

$$v = \frac{216}{R}$$

would serve, and we might allow slack gauge as follows:—

R	v
346 feet	$\frac{5}{8}$ inch
432 ,,	$\frac{1}{2}$,,
576 ,,	$\frac{3}{8}$,,
864 ,,	$\frac{1}{4}$,,

curves of more than 1000 feet radius being laid to straight gauge.

Double- or bull-headed rails are more easily adapted to ordinary curves than flat-footed rails, and for sharp curves the latter should be cautiously heated and bent to an extent which may be calculated by the same formula—

$$v = \frac{3\,C^2}{2\,R} \text{ (approximately)}.$$

In this case C represents the length of the rail of course.

Some platelayers break joint with flat-footed rails on sharp curves, to prevent the sharp elbows which are so often seen when the joints are opposite.

The next matter to consider is the cant or super-elevation of the outer rail. Avoiding the unscientific term "centrifugal force" in this connection—for the tendency of the train is not to fly outwards from the centre, but to move in a straight path—it is clear that, to change the direction of its movement so as to follow a curve, it is necessary to introduce a new force, always acting at right angles to the straight line in which the train tends to move—in other words, a force normal to the curve. In doing this we are simply following Newton's first law of motion. The force of gravity is therefore induced to effect this change of direction by giving a cant to the rails, either entirely by elevating the outer rail, or partly also by depressing the inner rail.

Let
- E = this superelevation,
- V = maximum velocity of trains,
- R = radius of curve,
- G = gauge,
- g = accelerative effect of gravity.

Then
$$\frac{E\,g}{G} = \frac{V^2}{R}$$
$$\therefore E = \frac{G\,V^2}{g\,R} \quad \ldots \ldots \ldots (2)$$

Otherwise, the chord C may be determined, the versed sine of which on any curve will give the required cant. Substituting $\frac{C^2}{8E}$ for R in formula (2) we obtain

$$C = \tfrac{1}{2} V \sqrt{G} \text{ (approximately)}. \quad \ldots \quad (2a)$$

It is convenient, however, to express E in inches, V in miles per hour, and G, C and R in feet. Adapted to these units, the formulæ are modified as follows:

$$\frac{E}{12} = \frac{G\,V^2}{32 \cdot 2\,R} \times \left(\frac{5280}{3600}\right)^2$$
$$\therefore E = \frac{G\,V^2}{1 \cdot 25\,R} \text{ (approximately)}. \quad \ldots \quad (2b)$$

$$C = \tfrac{1}{2} V \sqrt{G} \times \frac{5280}{3600}$$
$$\therefore C = 1\tfrac{1}{5} V \sqrt{G} \text{ (approximately)}; \quad \ldots (2c)$$

and, from the latter, we deduce the following formulæ:

$$\left.\begin{array}{ll}(5'\ 6''\text{ gauge}) & C = 1\tfrac{3}{4} V \\ (\text{metre gauge}) & C = 1\tfrac{1}{3} V \\ (4'\ 8\tfrac{1}{2}''\text{ gauge}) & C = 1\tfrac{3}{5} V\end{array}\right\} \quad \ldots (2d)$$
$$(\text{approximately}).$$

C having been thus calculated, a string of this length is stretched from one point to another on the inside of the curve, and its distance from the rail at the middle will be equal to the required cant.

At the end of the book is a table, giving the cant in inches for various speeds and curves, and also the chords, the versed sines of which are equal to the cant required for different speeds.

Should the cant calculated in this way prove, on trial, to be insufficient, the platelayer may lift the outer rail little by little until he is satisfied that engines run easily round the curve. The tendency is, however, to put on too great a cant, and platelayers should recollect that no amount of cant will prevent the sliding of the wheels on the outer rails, which must occur.

The elevation of the outer rail should begin some distance back on the straight, so that the full cant is attained at the beginning of the curve; and in the same way it is worked off on the straight at the other end of the curve.

In fact, the curve itself does not (or should not) end abruptly, as set out; but is gradually eased off into the straight, the curvature and cant diminishing together.

On sharp curves with flat-footed rails, check rails and a slower speed are preferable to too great a cant, since the tendency is to throw the weight of the train too much on the inner rail in the latter case and to draw the spikes.

53.—Laying Points and Crossings.

The more obtuse crossings, having an inclination of 1 in 5 or 1 in 6, are only used in temporary sidings or sidings to which engines are not admitted, except when such a crossing is laid with two curves of contrary flexure, and forms the junction between two split lines, as in Fig. 3. In this latter case, supposing the arrangement to be symmetrical, the radius of the diverging curves is equal to twice the radius of the curve which is put in for the same crossing off the straight, as in Fig. 2. Thus on the 5' 6" gauge a 1 in 6 crossing forming the junction between two split lines (Fig. 3) would be as easy to run over as a 1 in 8½ crossing laid in for a siding leaving a straight main line (Fig. 2), the radius of the curves being 800 feet in either case.

Medium crossings, having an inclination of 1 in 7 to 1 in 10, are used for ordinary work in station yards.

The more acute crossings, from 1 in 10 to 1 in 12, are required for facing points on main line junctions and similar positions, where an easy curve is of more importance than economy of space.

The lead of a crossing is the distance measured on the straight, from the heel of the switch to the intersection of the gauge-lines on the crossing. It is generally calculated with reference to that portion of a *uniform* curve which is intercepted between the heel of the switch and the nose of the crossing.

It is by no means necessary, however, to have a uniform curve from heel to nose. If a longer lead be put in, the curve will be flatter at the heel but sharper at the nose—and this, of course, is the worst possible arrangement; the idea of some platelayers that a crossing can be improved by putting in a long lead is utterly absurd. If, on the other hand, a shorter lead be adopted, we shall have a sharper curve near the heel, but we may also have a piece of straight over the crossing itself; so that this is the best arrangement of all when a sufficiently acute crossing is chosen to admit of the curve being made somewhat sharper at the heel.

The most common arrangement is that shown in Fig. 2, where one line turns out with a curve from a straight track, and the manner of laying points and crossings in this way will now be explained.

First of all the straight line is spiked with its stock-rail and switch in position. The opposite stock-rail and switch can then be placed to exact gauge, and the heel-chair spiked down. The siding stock-rail should be bent in slightly, just beyond the toe of the switch. While the jim-crow is still applied to the rail, the chair at the toe of the switch, the one in front, and the next slide-chair behind, should be spiked to somewhat tight gauge, and afterwards the rest of the slide-chairs. The tightness of gauge near the toe of the switch will disappear when the first train has passed over the crossing.

Similarly, in putting in the crossing, the chairs or lugs of the crossing on the straight line side should first be spiked, and then those on the other side.

The gauge should be, if anything, a little tight in front of the points, slack on the curve between the heel of the switch and the nose of the crossing, and exact at the nose of the crossing.

The cant, obtainable only by lowering the off siding rail, should begin at the butt of the stock-rail, and (unless there is a piece of straight over the crossing) be continued right through. This advice—to allow cant at the crossing—is by no means orthodox, but will be justified by trial and experience. To get a cant here the off rail must be countersunk, and it will be difficult to obtain more than an inch, but even this will materially assist the wheels over the curve and prevent them from mounting the nose of the crossing.

It is obvious, however, that a cross-over road must be on one dead level, for there is no room to run out the opposite cants before they meet midway.

The check-rails should be placed well forward to protect the nose of the crossing, and not so far back as to make one think that their purpose was forgotten.

Too many long sleepers should not be used. Three or four will be required under the crossing and two at the end of the switches to carry the lever, and these should be of sound hard wood, such as teak or sâl. Beyond the crossing the short sleepers of each road should be so laid alternately that they may easily be packed together, and that the dividing roads may be independently adjustable.

In all station yards spare point-rods, chairs, and switches should be kept ready for immediate renewals.

54.—Level Crossings.

To keep the wheel flange passages clear a check-rail is laid on the inside of each rail, and the space between should be cleared of dust and rubbish by the gateman, who ought also to water the crossing before a train is expected, to prevent dust from getting into the working parts of the engine. Sometimes, and with advantage, a check-rail is laid on both sides of the main rail.

55.—Final Remarks.

The amount of renewals annually required will probably vary in the following way.

During the first three or four years the annual renewals will diminish as the defective material is gradually weeded out. During a second period the renewals will be comparatively light for several years, and conform more or less with a general average. Then, as the term of life of one or other description of material is approached, very heavy extraordinary renewals will from time to time be required.

While, therefore, locomotive and traffic expenses vary generally with the expansion or contraction of traffic, the cost of maintaining the permanent way will occasionally appear to be irregular and excessive. This cannot always be taken as a sign of imperfect maintenance on the part of the engineers and platelayers. It is more often due to starving the renewals when the revenues of the railway are low, and allowing the defects to accumulate until heavy special renewals are suddenly and imperatively needed.

The statistics of renewals during any one year are seldom significant taken by themselves. Moreover, the distinction between special and ordinary renewals is fallacious; for the former are, as explained above, an accumulation of renewals which should have been regularly carried out in the ordinary course of maintenance. To enable one to draw sound inferences and thus make fairly accurate forecasts of annual requirements, the percentages of renewals up to date, ordinary and special, without distinction, should be added up, and an average struck at the end of each year.

CHAPTER III.

POINTS AND CROSSINGS.

56.—Object of Points and Crossings.

IT has already been explained that the use of points and crossings is to enable trains to pass from one pair of metals to another; that the switches are short tapering steel rails, one pressing against one stock-rail to divert the train from the original line, and the other drawn away from the other stock-rail to allow the train to pass on that side; and that the main line rail is crossed by means of a crossing (of cast steel, or built up of steel rails), the nose of the crossing being at the intersection of the gauge-lines of the crossing rails, which here diverge to form protecting wings.

57.—Definitions.

Gauge.—The distance between the rails of a track, measured from inside edge to inside edge of rail-heads. To measure the gauge from centre to centre of rail would quite unnecessarily introduce a new factor (the thickness of the rail-head), and complicate the formulæ to be investigated without the slightest reason for doing so.

Nose of crossing.—The point of intersection of the gauge-lines—an imaginary point, some inches in front of the blunt nose, to which all lead-measurements are referred.

Radius of crossing.—The radius of the curve of the gauge-line of one of the intersecting rails, the other being straight.

Clearance.—The clearance at heel of switch is the distance from inside edge of stock-rail to inside edge of switch-rail; in other words, the gap between those rails plus the breadth of the head of the rail.

Inclination-number, or number (to put it shortly) of the crossing. —The cotangent of the angle of the crossing; it is measured by the distance from nose of crossing at which the perpendicular offset from one gauge-line of the crossing to the other is unity.

Curve-lead.—The distance from springing of curve to nose of crossing, measured along the straight.

Switch-lead.—The distance from springing of curve to heel of switch, measured along the straight.

Lead of crossing.—The distance from nose of crossing to heel of switch, measured along the straight. By one method of calculation it is equal to the curve-lead minus the switch-lead.

58.—Symbols.

G = gauge.
a = angle of crossing.
$I = \cot a$ = number of crossing.
l = length of switch.
d = clearance.
$\delta = \sin^{-1}\left(\dfrac{d}{l}\right)$ = angle of divergence of switch.
$i = \dfrac{l}{d}$ = inclination-number of the switch.
R, R_1, R_2 = radii of curves of intersecting rails.
L' = curve-lead.
l' = switch-lead.
L = lead of crossing.

59.—Two Methods of determining the Lead.

For the purpose of primary investigation the simplest case may be considered—that of a turnout from a straight line—but it will afterwards be seen that the application to cases where both lines are curved is easy.

In determining the lead by the first method, the curve is assumed to lie tangentially upon the gauge-line of the switch-rail, springing from it at the heel and crossing the inner straight rail at the proper angle. The curve is not referred in any way to the gauge-line of the outer rail. We do not take it for granted that a switch will be used which will suit a previously determined curve, but accept a certain switch, clearance, and crossing, and make the lead and curve suit those conditions.

By the second method the first step is to determine the radius of the curve which would lie tangentially upon the gauge-line of the outer straight rail and cross the inner straight rail at the proper angle. The curve-lead may then be found, and also the switch-lead, the latter covering that portion of the curve which (being intercepted between the heel of the switch and the theoretical springing point on the gauge-line of the outer straight rail) is imaginary. The lead of the crossing is simply the difference between the curve-lead and the switch-lead.

Of these two methods, the first is more immediately suggested by what occurs in practice. If the lines were laid according to the second method there would be a kink at the heel unless the switch were of a certain length. Now, as a matter of convenience, the same switch is often used for different crossings and cannot possibly be correct for more than one. The curve, therefore, requires generally a little adjustment to avoid a kink at the heel of the switch. This cannot occur when the first method is adopted, since the given switch is at once made a datum-line from which the curve starts. As, however, it is convenient to have certain leads laid down for certain crossings without reference to the length of the switch, the second method is usually followed, and the slight error due to using a switch whose length is not that of the tangent to the imaginary portion of the curve may practically be ignored.

60.—First Method. Lead of Crossing.

Referring to figure 1,

$$\delta = \sin^{-1}\left(\frac{d}{l}\right)$$

$$i = \frac{l}{d}$$

as defined.

Then,

$$\angle CED = \angle KCE$$
$$= \angle KCF + \angle FCE$$
$$= \angle KCF + \angle FEC$$
$$= a + \delta - \angle CED$$
$$\therefore \angle CED = \frac{a + \delta}{2}$$

Moreover,

$$L = DE$$
$$= CD \cot CED$$
$$\therefore L = (G - d) \cot\left(\frac{a + \delta}{2}\right) \quad \ldots \quad (3)$$

Hence,

$$L = (G - d)\frac{\cot\frac{a}{2} \cot\frac{\delta}{2} - 1}{\cot\frac{a}{2} + \cot\frac{\delta}{2}}$$

$$= (G - d)\frac{4Ii - 1}{2(I + i)} \text{ (approximately)}$$

$$\left.\begin{array}{l}\therefore L = 2(G - d)\dfrac{Ii}{I + i} \text{ (approximately)} \\ = 2(G - d)\dfrac{Il}{Id + l} \text{ (approximately)}.\end{array}\right\} \quad (3a)$$

61.—First Method. Radius of Crossing.

Again, referring to figure 1,

$$CE = \frac{G - d}{\sin\left(\dfrac{a + \delta}{2}\right)}$$

But,
$$R = \frac{CE}{2} \sin\left(\frac{a-\delta}{2}\right)$$

$$\therefore R = \tfrac{1}{2}(G-d) \frac{\sin\left(\frac{a-\delta}{2}\right)}{\sin\left(\frac{a+\delta}{2}\right)} \quad \ldots \quad (4)$$

62.—Second Method. Radius of Crossing.

Referring to figure 2,
$$R = \frac{G}{\text{versin } a} \quad \ldots \ldots \ldots (5)$$

This will be a convenient formula if the crossing is described in terms of the angle, and that angle is simply expressed. As, however, R is found by dividing a small number by a *very* small number, the slightest roughness of approximation in the denominator may affect the result considerably. Hence, if the crossing is described in terms of the inclination — 1 in 7, 1 in 8, and so on—instead of first deducing the angle of the crossing in order to use the above, it will be better to recast the formula at once in terms of G and I :

$$\cot a = I$$
$$\therefore \cos a = \frac{I}{\sqrt{(1+I^2)}}$$
$$R = \frac{G}{1 - \cos a}$$
$$\left.\begin{aligned}\therefore R &= G(1+I^2) + GI\sqrt{(1+I^2)} \\ &= 2GI^2 \text{ (approximately)} \\ &= 2GI^2 + 1\tfrac{1}{2}G \text{ (very closely).}\end{aligned}\right\} \quad \ldots (5a)$$

63.—Curve-lead.

Again referring to figure 2,
$$L' = R \sin a \quad \ldots \ldots \ldots (6)$$

and, by easy transpositions,

$$L' = \frac{R}{\sqrt{(1+I^2)}} \atop = \frac{R}{I} \text{ (approximately)} \right\} \quad \ldots \ldots (6a)$$

or, in terms of G and I,

$$L' = G\sqrt{(1+I^2)} + GI \atop = 2GI \text{(approximately)} \right\} \quad \ldots \quad (7)$$

or, again, in terms of G and a,

$$L' = G \cot \frac{a}{2} \quad \ldots \ldots \ldots (7a)$$

moreover,

$$2R - G : L' :: L' : G;$$
$$\therefore L' = \sqrt{(2R - G)G} \atop = \sqrt{2RG} \text{ (approx.)} \right\} \quad \ldots \quad (8)$$

64.—Switch-lead.

Similarly we obtain the following formulæ for the switch-lead:—

$$l' = \sqrt{(2R - d)d} \atop = \sqrt{2Rd} \text{ (approx.)} \right\} \quad \ldots \ldots \ldots (9)$$

$$l' = \sqrt{2Gd\{(1+I^2) + I\sqrt{(1+I^2)}\} - d^2} \atop = 2I\sqrt{Gd} \text{ (approx.)} \right\} \quad (10)$$

65.—Second Method. Lead of Crossing.

To find the lead of the crossing we have merely to subtract the switch-lead from the curve-lead:

$$L = L' - l' \quad \ldots \ldots \ldots (11)$$

66.—The Platelayer's Rule for the Lead.

The approximate formula for the curve-lead,

$$L' = 2GI,$$

gives us the well-known "platelayer's rule" for the lead:—
"The lead is equal to twice the gauge multiplied by the number of the crossing." This rule, however, does not give the lead of the crossing but only the curve-lead from which the switch-lead has still to be deducted.

The following formulæ are more nearly correct:

$$(5' \; 6'' \text{ gauge}) \; L = 1\tfrac{1}{2} \, G \, I,$$
$$(\text{metre gauge}) \; L = 1\tfrac{2}{3} \, G \, I,$$
$$(4' \; 8\tfrac{1}{2}'' \text{ gauge}) \; L = 1\tfrac{3}{7} \, G \, I;$$

but still simpler empirical formulæ for calculating the lead may be given here:

$$\left. \begin{array}{l} (5' \; 6'' \text{ gauge}) \; L = 8\tfrac{1}{4} \, I \\ (\text{metre gauge}) \; L = 4\tfrac{3}{5} \, I \\ (4' \; 8\tfrac{1}{2}'' \text{ gauge}) \; L = 6\tfrac{4}{5} \, I \\ \quad (\text{approximately}). \end{array} \right\} \quad \ldots \quad (12)$$

These may be deduced from the approximate formula,

$$L = 2I \, (G - \sqrt{G \, d}).$$

67.—Curves of Contrary Flexure.

In figure 3,

$$BD^2 - CD^2 = AB^2 - AC^2$$

$$\therefore BD - CD = \frac{AB^2 - AC^2}{BC}$$

$$= \frac{R_1^2 - R_2^2}{R_1 + R_2 - G}.$$

Again,

$$BD = \tfrac{1}{2}(BC + \overline{BD - CD})$$

$$= \tfrac{1}{2}\left(R_1 + R_2 - G + \frac{R_1^2 - R_2^2}{R_1 + R_2 - G} \right).$$

But

$$AD = \sqrt{AB^2 - BD^2}$$

$$\therefore L' = \sqrt{R_1^2 - \tfrac{1}{4}\left(R_1 + R_2 - G + \frac{R_1^2 - R_2^2}{R_1 + R_2 - G} \right)^2} \quad (13)$$

An approximate formula which will be much more serviceable can, however, easily be found.

If s and $G - s$ be the cross-distances from the nose of the crossing to the lines tangential to the curves at springing

$$L' = \sqrt{2 R_1 s}; \text{ and also}$$
$$L' = \sqrt{2 R_2 (G - s)} \text{ (approximately).}$$

Whence, by eliminating s,

$$L' = \sqrt{\frac{2 R_1 R_2 G}{R_1 + R_2}} \text{ (approximately),} \quad . \quad . \quad . \quad (13a)$$

and, in the same way,

$$l' = \sqrt{\frac{2 R_1 R_2 d}{R_1 + R_2}} \text{ (approximately).} \quad . \quad . \quad . \quad (14)$$

When the radii of the curves are equal, we obtain the following formulæ:

$$R = \frac{G}{2 \text{ versin } \frac{a}{2}} \quad . \quad . \quad . \quad . \quad . \quad . \quad . \quad . \quad (15)$$

$$= 4 G I^2 \text{ (approximately)} \quad . \quad . \quad . \quad . \quad (15a)$$

$$L' = R \sin \frac{a}{2} . \quad . \quad . \quad . \quad . \quad . \quad . \quad . \quad . \quad (16)$$

$$= \frac{G}{2} \cot \frac{a}{4} . \quad . \quad . \quad . \quad . \quad . \quad . \quad . \quad . \quad (17)$$

$$= 2 G I \text{ (approximately)} \quad . \quad . \quad . \quad (17a)$$

$$= \sqrt{\left(R - \frac{G}{4}\right) G} . \quad . \quad . \quad . \quad . \quad . \quad (18)$$

$$= \sqrt{R G} \text{ (approximately).} \quad . \quad . \quad . \quad (18a)$$

$$l' = \sqrt{R d} \text{ (approximately)} \quad . \quad . \quad . \quad . \quad (19)$$

$$= 2 I \sqrt{G d} \text{ (approximately).} \quad . \quad . \quad (20)$$

68.—Curves of Similar Flexure.

In figure 4
$$BD^2 - CD^2 = AB^2 - AC^2$$
$$\therefore BD + CD = \frac{AB^2 - AC^2}{BC}$$
$$= \frac{R_1^2 - R_2^2}{R_1 - R_2 - G}$$

and, finally, by a process similar to that by which the case of curves of contrary flexure was determined,

$$L' = \sqrt{R_1^2 + \tfrac{1}{4}\left(R_1 - R_2 - G + \frac{R_1^2 - R_2^2}{R_1 - R_2 - G}\right)^2} \quad (21)$$

Again, if s and $G + s$ be the cross-distances from the nose of the crossing to the lines tangential to the curves at springing

$$L' = \sqrt{\frac{2 R_1 R_2 G}{R_1 - R_2}} \text{ (approximately)} \quad . \quad . \quad (21a)$$

$$l' = \sqrt{\frac{2 R_1 R_2 d}{R_1 - R_2}} \text{ (approximately)}. \quad . \quad . \quad (22)$$

69.—Lead of Crossing Constant.

The formulæ for leads, thus obtained, in the case of curves of similar or contrary flexure are, however, of theoretical rather than practical interest.

Once having found the crossing which will suit a turnout from a straight line or a turnout from another curve of similar or contrary flexure, the lead as worked out by the formulæ applicable to each case, will practically be the same.

Thus, on the 5′ 6″ gauge, whether we put in a 7° 45′ crossing for two curves of contrary flexure, their radii being 1200 feet, or for a turnout from a straight line with a curve of 600 feet, the lead will be 60′ 7″ in either case.

This simplifies matters greatly. We have merely to find

what crossing is suitable and to determine the lead for the simplest case—that of a turnout from a straight line. Then this lead, provided the crossing be suitable, will be the same for that crossing, whatever the arrangement may be.

It is also to be remarked that when the radii of two curves of contrary flexure are equal, each is very nearly equal to twice the radius of the crossing, that is, twice the radius of the curve of a turnout from a straight line.

70.—To Determine the Crossing.

1. Turnout with a curve of given radius from a straight line. (Fig. 2.)

In this case we need only express formula (5a) in another form:—

$$\left. \begin{array}{l} I = \dfrac{R - G}{\sqrt{(2R - G)G}} \\ = \sqrt{\dfrac{R}{2G}} \text{ (approximately),} \end{array} \right\} \quad \ldots \ (23)$$

while formula 5 becomes

$$\cos a = \frac{R - G}{R} \quad \ldots \ldots \ldots \ldots (23a)$$

by which the angle of crossing may be determined.

2. Curves of contrary flexure. (Fig. 3.)

Since the lead is practically constant, whatever may be the arrangement

$$2GI = \sqrt{\frac{2R_1 R_2 G}{R_1 + R_2}}$$

$$\therefore I = \sqrt{\frac{R_1 R_2}{2(R_1 + R_2)G}} \text{ (approximately); } \quad (24)$$

and, when the radii are equal—

$$I = \sqrt{\frac{R}{4G}} \text{ (approximately).} \quad \ldots \ldots (25)$$

3. *Curves of similar flexure.* (Fig. 4.)

Here, in a similar manner, we obtain—

$$I = \sqrt{\frac{R_1 R_2}{2(R_1 - R_2)G}} \text{ (approximately)}. \quad \ldots \quad (26)$$

71.—Modification of the Lead in certain cases.

The inclination of a crossing is defined by such short lines that a slight difference from the accurately calculated lead may be permitted in order to avoid cutting rails as much as possible.

Hence also, if it be necessary to put in a crossing with a curve or curves whose radii differ somewhat from those which exactly suit the crossing, the lead may be adapted to those curves by means of the formulæ which give the lead in terms of the radii and the gauge; or the following approximate formulæ may be applied:

$$L = K\sqrt{R} \text{ (approximately)} \quad \ldots \quad (27)$$

for a turnout from a straight line;

$$L = K\sqrt{\frac{R_1 R_2}{R_1 + R_2}} \text{ (approximately)} \quad \ldots \quad (28)$$

for curves of contrary flexure; and

$$L = K\sqrt{\frac{R_1 R_2}{R_1 - R_2}} \text{ (approximately)} \quad \ldots \quad (29)$$

for curves of similar flexure; where

$$\left.\begin{array}{l} K = 2\tfrac{1}{2} \text{ (5' 6" gauge)} \\ K = 1\tfrac{1}{5} \text{ (metre gauge)} \\ K = 2\tfrac{1}{5} \text{ (4' 8}\tfrac{1}{2}\text{" gauge)}. \end{array}\right\}$$

This concession, however, is no argument for indifference to accuracy in calculation. In practice we may find it convenient to depart from the correct dimensions, but we ought to know exactly to what extent we are doing so.

Some platelayers have an idea that they can make a crossing easier by lengthening the lead. This is absurd. They simply flatten the curve near the switch, and sharpen the curve near the crossing, i. e. at the very worst place.

In fact, if the curve be easy, it may be advisable to shorten the lead; for although, in this case, the curve would have to be sharper towards the switch, we might be able to put in a piece of straight at the crossing, which would be an improvement.

So that we may say generally that the best arrangement is when an easy or acute-angled crossing is put in with a short lead (which may easily be calculated for any desired smaller radius of curve), and with a piece of straight over the crossing.

72.—Turnouts and Crossovers.

A turnout from one line to another (figures 5 and 7) is made by means of a pair of switches, a crossing, and a reverse curve running into the direction of the second line, with or without an intermediate portion of straight line.

When a turnout enters the branch-line by means of a second crossing, and a second pair of switches, it becomes a crossover (figures 6 and 8).

The whole length of a turnout, therefore, is made up of

(a) Lead of crossing (L),
(b) Intermediate portion (S), straight or curved,
(c) Curve-lead (L');

while that of a crossover is made up of

(a) Lead of crossing (L),
(b) Intermediate portion (S), straight or curved,
(c) Lead of crossing (L).

1. Turnouts and crossovers, the intermediate portion being straight.

Here (Figs. 5 and 6) we have

$$S = \{D - G(1 + \sec a)\} \cot a \\ = (D - 2G) \, l \text{ (approximately)} \qquad \ldots \quad (30)$$

2. Turnouts and crossovers, the intermediate portion being curved (Figs. 7 and 8).

The curve is continuous and reverse, the radius R in the following formula being that of the crossing rail :

$$S = \sqrt{D(4R - 2G - D)} - 2L'. \quad . \quad . \quad (31)$$

73.—Gathering Lines.

In Fig. 9, let

S = distance from nose to nose of crossings measured along the lines.

V = the same measured along the gathering line.

D = distance between tracks.

α = angle of crossings.

θ = angle which the gathering line makes with the parallel lines.

Then

$$V = D \operatorname{cosec} \theta. \quad . \quad . \quad . \quad . \quad . \quad (32)$$
$$S = D \cot \theta. \quad . \quad . \quad . \quad . \quad . \quad . \quad (33)$$

and, to find the position of the crossings,

$$T = R \tan\left(\frac{\theta - \alpha}{2}\right). \quad . \quad . \quad . \quad . \quad (34)$$

$$Y = \frac{T \sin \alpha}{\sin \theta}. \quad . \quad . \quad . \quad . \quad . \quad . \quad (35)$$

$$Z = \frac{T \sin (\theta - \alpha)}{\sin \theta}. \quad . \quad . \quad . \quad . \quad (36)$$

When θ = "the limiting angle," i.e. the greatest angle at which the gathering line can be run across the parallel lines, each crossing fits against the butt of the next stock-rail; no piece is put in, and no room is lost between them; if the gathering line be indefinitely extended, the tracks being equidistant, each crossing in succession will occupy a similar position; and there will be no fear of being jammed.

The examination of this case will be of assistance where

economy of space has to be studied in laying out a yard, and will prevent the designer, perhaps, from proposing impossible arrangements in the plan.

To find θ, the limiting angle, V must be equal to the sum of the length of the crossing, the lead, the switch, and of that part of the stock-rail which lies beyond the toe of the switch and butts against the crossing, while D must be equal to the least distance between tracks.

Thus, with 7° 45′ crossings on the metre gauge, $V = 4' + 33' \ 9'' + 15' \ 1'' = 52' \ 10''$; $D = 12$; $\sin \theta = \dfrac{12}{52 \cdot 8333}$; and θ, the limiting angle, $= 13° \ 7' \ 41''$.

When the gathering line runs across the parallel lines at the angle of the crossings as in figure 10, $\theta = a$, and

$$V = D \operatorname{cosec} a \quad \ldots \ldots \ldots \quad (37)$$
$$S = D \cot a \quad \ldots \ldots \ldots \quad (38)$$

In this case we may, if we wish, put in compounds within the crossings as in figures 11 and 12, and the parallel lines and gathering line may cross one another. Thus in figures 11 and 12

$$T = G \operatorname{cosec} a \quad \ldots \ldots \ldots \quad (39)$$

$$R = G \cot \frac{a}{2} \quad \ldots \ldots \ldots \quad (40)$$

$$Y = R \left(\sec \frac{a}{2} - 1\right) \quad \ldots \ldots \ldots \quad (41)$$

$$\left. \begin{array}{l} l' = \sqrt{(2R - d)d} \\ = \sqrt{2Rd} \ (\text{approximately}) \end{array} \right\} \quad \ldots \quad (42)$$

74.—Three-throw Points and Crossings.

Here (figure 13) the curve-lead is the same, but the switch-lead must be calculated for a double clearance:

$$\left. \begin{array}{l} l' = \sqrt{4d(R - d)} \\ = 2\sqrt{Rd} \ (\text{approximately}) \end{array} \right\} \quad \ldots \quad (43)$$

The lead, therefore, of the centre crossing is less by the

greater length of the switch-lead; while that of each outer crossing is the same as usual.

We may note, in passing, that such an arrangement may be made with a 1 in 6 and two 1 in 8½ crossings on the 5′ 6″ gauge, the radius being 800 feet; and with a 1 in 6¾ and two 1 in 9½ crossings on the metre gauge, the radius being 600 feet; or, with a 10° 45′ crossing and two 7° 45′ crossings on the 5′ 6″ gauge, the radius being 600 feet; and with an 11° 0′ crossing and two 7° 45′ crossings on the metre gauge, the radius being 360 feet.

75.—Triangles.

At temporary termini, where turntables will not ultimately be required, it is usual to put in triangles for reversing engines.

One arrangement of triangle is that shown in Fig. 14. So far as the purpose of turning the engine is concerned, the third side A B is not necessary, but it generally forms part of an already existing through siding.

The distance from the theoretical springing of the curve of crossing A to that of crossing B is obviously equal to $2R - G$, where R is the radius of the crossing rails; and the perpendicular distance from the theoretical springing of crossing C to the centre of the track A B is equal to $R - \dfrac{G}{2}$.

Hence, if S = distance between the crossings A and B, and L'_1 = curve-lead of each of those crossings,

$$S = 2R - G - 2L'_1 \quad \ldots \quad (44)$$

and if D = distance of crossing C from the centre-line of the track A B, and L'_2 the curve-lead,

$$D = R - \frac{G}{2} - L'_2 \quad \ldots \quad (45)$$

The crossings A and B have the same angle, while that of C is chosen to suit the particular radius.

PLATELAYING, AND POINTS AND CROSSINGS. 73

Thus, on the 5' 6" gauge, 1 in 8½ crossings may be placed at A and B, and a 1 in 6 at C. Moreover,

R = 802·99; S = 1,412'10"; and D = 733'10".

If a triangle be laid out as shown in figure 15, it occupies the least possible space for a given limiting radius. The arrangement is equiangular, each crossing having the same angle.

Let R = radius of the crossing rails; S = distance between the crossings; and L' = curve-lead of each of the crossings.

Then, evidently,

$$S = R - L'\sqrt{3} \quad \ldots \ldots \quad (46)$$

Thus, on the 5' 6" gauge, with a 1 in 6 crossing at A, B, and C, R = 800; L' = 66' 5"; S = 685.

76.—Crossing more than one Line; Diamond Crossings.

Referring to figure 16, in order to obtain the curve-leads or distances from the theoretical springing of the curve on the first line, we must substitute for G and R in the formula for curve-lead,

$$L' = \sqrt{(2R - G)G},$$

the following:—

for No. 1 crossing .. .. G and R,
for No. 2 crossing .. .. D − G and R − G,
for No. 3 crossing .. .. D + G and R,
for No. 4 crossing .. .. 2D − G and R − G,
for No. 5 crossing .. .. 2D + G and R.

Thus we have—

$$\left. \begin{array}{l} L'_1 = \sqrt{(2R - G)G} \\ L'_2 = \sqrt{(2R - G - D)(D - G)} \\ L'_3 = \sqrt{(2R - G - D)(D + G)} \\ L'_4 = \sqrt{(2R - G - 2D)(2D - G)} \\ L'_5 = \sqrt{(2R - G - 2D)(2D + G)} \end{array} \right\} \quad \ldots \quad (47)$$

and so on.

To determine the crossings:

$$\left.\begin{aligned} \sin a_1 &= \frac{L'_1}{R} \\ \sin a_2 &= \frac{L'_2}{R-G} \\ \sin a_3 &= \frac{L'_3}{R} \\ \sin a_4 &= \frac{L'_4}{R-G} \\ \sin a_5 &= \frac{L'_5}{R} \end{aligned}\right\} \quad \ldots \ldots (48)$$

The case of crossings on mixed gauges is similar, and need not be more particularly described.

77.—Tables.

In addition to tables of dimensions for points and crossings, turnouts, and crossovers for the 5' 6", the metre, and the 4' 8½" gauges (including special tables for the old and new standard crossings in use on Indian State railways), a graphic diagram is given at the end of the book, but, the scale being a small one, more for the sake of example than use. The universal application and concise completeness of such graphic diagrams, and their practical superiority over tables of figures are obvious enough; and the object of the table of numbers, angles, radii, curve-leads, switch leads, and leads for crossings for each gauge is to enable the student to plot such graphic diagrams to a larger scale for his own use.

CHAPTER IV.

SIMPLE RULES.

78.—To Set out a Curve by Offsets.

Rule.—To find O, the full offset for 100 feet chords, divide 10,000 by the radius R (figure 17).

Formula.—
$$O = \frac{10,000}{R} \quad \ldots \ldots \quad (49)$$

Proof.—By similar triangles in figure 17.
$$\frac{O}{C} = \frac{C}{R}$$
$$\therefore O = \frac{C^2}{R}$$
$$\therefore O = \frac{10,000}{R}$$

Directions.—Having placed a flag at the tangent peg, measure 100 feet along the tangent and place a second flag. From the latter, at right angles to the tangent, measure half the offset O, i.e. $\frac{O}{2}$, and drive a peg at the point thus found, the second point on the curve.

From here, in line with the tangent and second pegs, measure 100 feet and place a flag. Then shift the chain until the leader's end is at a distance equal to the offset O from the flag, while the follower's end is still held at the second peg. Drive a peg at the leader's end of the chain. This will be the third point on the curve.

The fourth and succeeding points may be found in the same way as the third, until the end of the curve is reached.

Here the direction of the tangent is determined by laying off the half-offset, $\dfrac{O}{2}$, instead of the full offset, O.

Example.—Let R = 1273·2. Then,

$$O = \dfrac{10,000}{1273 \cdot 2}$$

$$= 8 \cdot 1449$$

$$\therefore O = 8' \, 1\tfrac{3}{4}''.$$

79.—To Set out a Diversion from a Straight Line.

Rule.—To find L, the length of each half of the diversion (Fig. 18), add together the square (S^2) of the straight portion between the reverse curves and the product (4 R D) of four times the radius into the maximum distance of the diverted from the main line; subtract the square (D^2) of the maximum distance; and take the square root of the result.

To find the tangent (T), divide the product (R D) of the radius into the maximum distance by the sum (L + S) of the half of the diversion and the length of straight between the reverse curves.

Formulæ.—

$$L = \sqrt{S^2 + 4RD - D^2} \quad \ldots \quad (50)$$

$$T = \dfrac{RD}{L + S} \quad \ldots \ldots \ldots (51)$$

Proof.—By similar triangles in figure 18,

$$\dfrac{T}{R} = \dfrac{D}{L + S}$$

$$\therefore T = \dfrac{RD}{L + S}$$

Again,
$$L = 2T + \sqrt{(2T + S)^2 - D^2}$$
$$\therefore (L - 2T)^2 = (2T + S)^2 - D^2$$
$$\therefore L^2 = 4T(L + S) + S^2 - D^2$$
$$\therefore L^2 = \frac{4RD}{L + S}(L + S) + S^2 - D^2$$
$$\therefore L = \sqrt{S^2 + 4RD - D^2}.$$

Directions.—From the middle point of the line, which is to be diverted, set off the maximum offset D. From the middle point of the diversion thus determined, measure parallel to the main line and in both directions the tangent lengths T. This fixes on each side one extremity of the common tangent lines.

From the middle point of the line, which is to be diverted, measure along the line in both directions the half length L. This determines the beginning and the end of the diversion, from each of which again the distance T must be measured back along the line, to fix on each side the other extremity of the common tangent lines.

These may now be lined in, and the four tangent lengths T measured off, leaving the required length of straight S in the middle.

We have now fixed all the points that we require, and the line of the diversion may be laid in accordingly.

If necessary, a piece of straight may be allowed for at the middle of the diversion by making the whole length of the diversion so much longer; but, unless a long piece of straight is required, it is better to make the middle curve continuous, as shown in the figure, to avoid having to work off and resume the superelevation of the outer rail on the curves.

Example.—Let R = 1000, S = 100, and D = 50. Then
$$L = \sqrt{100^2 + (4 \times 1000 \times 50) - 50^2}$$
$$\therefore L = 455\tfrac{1}{2}, \text{ and the whole length} = 911 \text{ feet.}$$

Also

$$T = \frac{1000 \times 50}{455\frac{1}{2} + 100}$$

$$\therefore T = 90 \text{ feet.}$$

80.—To Find the Radius of a Curve.

Rule.—Stretch a string 100 feet long upon the inside edge of the curved rails, and measure with a foot-rule the versine or distance from the middle of the string to the rail. Then the radius R is very nearly equal to 15,000 divided by v, the versine expressed in inches.

Formula.—

$$R = \frac{15,000}{v} \quad \ldots \ldots \ldots (1)$$

where v = versine in inches.

Example.—Let $v = 10$ inches. Then

$$R = \frac{15,000}{10}$$

$$\therefore R = 1500 \text{ feet.}$$

81.—To Find the Cant Required on a Curve.

Rule.—Stretch a string of a certain length, according to the gauge and maximum speeds of trains, upon the inside of the curve. Then the versine is equal to the required cant.

On the 5' 6" gauge the length of this string should be equal to one and three-quarter times V, where V is expressed in miles per hour.

On the metre gauge, the string should be one and one-third times V.

On the 4' 8½" gauge, the string should be one and three-fifths V.

See also the table of chords determining the cant, which is given at the end of the book.

Formulæ.—

$$\left.\begin{array}{l}(5'\ 6''\text{ gauge})\ C = 1\tfrac{3}{4}\ V \\ (\text{metre gauge})\ C = 1\tfrac{1}{3}\ V \\ (4'\ 8\tfrac{1}{2}''\text{ gauge})\ C = 1\tfrac{3}{5}\ V\end{array}\right\} \quad \ldots \quad (2d)$$

where C = the required length of string and V = the maximum speed of trains in miles per hour.

Example.—On the metre gauge, let V = 30 miles per hour. Then
$$C = 1\tfrac{1}{3} \times 30,$$
$$\therefore C = 40 \text{ feet.}$$

If, therefore, a string 40 feet long be stretched along a curve whose radius is 1000 feet, for example, we shall find that the versine is $2\tfrac{3}{8}$ inches, and this is the cant or superelevation which must be given to the outer rail.

82.—To find how much a Rail should be bent to suit a Curve.

Rule.—If C be the length of the rail, divide three times C^2 by twice the radius R, these being expressed in feet. The result will give v, the versine expressed in inches; and the rail must be curved until, if a string be stretched from end to end, the distance from the middle of the string to the rail is equal to v.

Formula.—
$$v = \frac{3\ C^2}{2\ R}, \quad \ldots \ldots \ldots (1)$$

where C and R are expressed in feet and v in inches.

Example.—Let C = 24 feet and R = 1200.
$$v = \frac{3 \times 24 \times 24}{2 \times 1200},$$
$$\therefore v = \tfrac{3}{4} \text{ inch.}$$

83.—To find the Number of a Crossing.

Rule.—If intersecting strings be laid along the gauge-lines or edges of the crossing, the number I of the crossing is measured by the distance in feet from the intersection of the strings beyond the nose to the point on one string whence the perpendicular offset to the other string is equal to 1 foot.

Example.—If the distance at which the offset is 1 foot be 9′ 9″ the crossing is called a "1 in $9\frac{3}{4}$," and the number of the crossing is 9·75, or $9\frac{3}{4}$.

84.—To find the Radius of a Crossing.

Rule.—Multiply I^2, or the square of the number of the crossing (Fig. 2), by twice the gauge, or 2 G. For a closer approximation add $1\frac{1}{2}$ G to the result.

Formula.—
$$R = 2\,G\,I^2 + 1\tfrac{1}{2}\,G \quad \ldots \ldots \quad (5a)$$

Example.—Let G = 5′ 6″ and I = $8\frac{1}{2}$. Then
$$R = (2 \times 5\tfrac{1}{2} \times 8\tfrac{1}{2} \times 8\tfrac{1}{2}) + 8\tfrac{1}{4},$$
$$\therefore\ R = 803.$$

Now the accurately calculated radius is 802·99, so that the approximation is wonderfully close.

To test the formula by another example, let G = 3′ $3\frac{3}{8}$″ and I = 12. Then,
$$R = (2 \times 3\cdot 2809 \times 12 \times 12) + (1\tfrac{1}{2} \times 3\cdot 2809);$$
$$\therefore\ R = 949\cdot 82,$$
which scarcely differs at all from 949·83, the actual radius.

85.—To find the Lead of a Crossing.

Rule.—On the 5′ 6″ gauge multiply the number of the crossing (Fig. 2) by eight and one quarter; on the 3′ $3\frac{3}{8}$″ gauge by four and three-fifths; and on the 4′ $8\frac{1}{2}$″ gauge by six and four-fifths.

PLATELAYING, AND POINTS AND CROSSINGS. 81

Formulæ.—
$$\left. \begin{array}{l} (5'\ 6''\ \text{gauge})\ L = 8\tfrac{1}{4}\,I \\ (\text{metre gauge})\ L = 4\tfrac{3}{5}\,I \\ (4'\ 8\tfrac{1}{2}''\ \text{gauge})\ L = 6\tfrac{4}{5}\,I \end{array} \right\} \quad \ldots \ldots \ (12)$$

where L = lead of crossing.

Example.—Let $G = 5'\ 6''$ and $I = 8\tfrac{1}{2}$. Then,
$$L = 8\tfrac{1}{4} \times 8\tfrac{1}{2};$$
$$\therefore\ L = 70\tfrac{1}{8}\ \text{feet.}$$

Now, the actual lead is 70, so that the approximation is sufficiently close for all ordinary purposes.

86.—To find the Crossing required in any Case.

(1). Turnout with a curve at a given radius from a straight line (Fig. 2).

Formula.—
$$I = \sqrt{\frac{R}{2\,G}} \cdot \quad \ldots \ldots \ (23)$$

Example.—On the 5' 6½" gauge let $R = 1100$. Then,
$$I = \sqrt{\frac{1100}{2 \times 5\tfrac{1}{2}}}$$
$$\therefore\ I = 10$$

that is, a 1 in 10 crossing may be used.

(2). Curves of contrary flexure (Fig. 3).

Formula.—
$$I = \sqrt{\frac{R_1\,R_2}{2\,(R_1 + R_2)\,G}} \cdot \quad \ldots \ldots \ (24)$$

where R_1 and R_2 are the radii of the curves, which turn off to right and left.

Example.—Let $G = 5'\ 6''$, $R_1 = 2475$, and $R_2 = 1980$. Then,
$$I = \sqrt{\frac{2475 \times 1980}{2 \times 4455 \times 5\tfrac{1}{2}}}$$
$$\therefore\ I = 10$$

that is, a 1 in 10 crossing will suit this case also.

(3). Curves of similar flexure (Fig. 4).

Formula.—
$$I = \sqrt{\frac{R_1 R_2}{2(R_1 - R_2) G}} \quad \ldots \ldots (26)$$

where R_1 and R_2 are the radii of the curves, which both turn off in the same direction.

Example.—Let $G = 5'\ 6''$, $R_1 = 1650$, and $R_2 = 660$. Then
$$I = \sqrt{\frac{1650 \times 660}{2 \times 990 \times 5\tfrac{1}{2}}}$$
$$\therefore I = 10$$

so that a 1 in 10 crossing will suit this case also, and the lead is in every case 82′ 3″.

87.—To Find the Distance from Nose to Nose of Crossings in a Crossover.

Rule.—When the intermediate portion is straight, the distance S (Fig. 6) from nose to nose of crossings measured along one of the straight main lines is equal to the number I of the crossing multiplied by the difference between D, the distance from centre to centre of main lines, and 2 G, twice the gauge.

Formula.—
$$S = I(D - 2G) \quad \ldots \ldots (30)$$

Example.—Let $G = 5'\ 6''$ and $D = 15$, each crossing being a 1 in $8\tfrac{1}{2}$. Then
$$S = 8\tfrac{1}{2}(15 - 11).$$
$$\therefore S = 34 \text{ feet.}$$

Now by the exact formula this would be 33′ 8″, so that the difference is insignificant.

PLATELAYING, AND POINTS AND CROSSINGS.

REPORT ON STEEL RAILS.

Weight of rails, 41¼ lbs. per yard.

Lengths allowed, 94 per cent. 24 ft. ... Weights 330·00 lbs.
" " 4 " 21 " ... " 288·75 "
" " 2 " 18 " ... " 247·50 "

Tests of Section.

Date.	No. of Rails Weighed.	Total Weight in lbs.	Average Weight of 30 ft. Rail.
March 17	6	1988	330·76 or 41·34 pounds per yard.
" 21	6	1981	
	12	3969	

Lever Test of 24 ft. Rails on 3 ft. Bearing.

Date.	10 Tons.	Perma-nent Set.	12 Tons.	Perma-nent Set.	16 Tons.	Perma-nent Set.	20 Tons.	Perma-nent Set.
March 30	⅜ ⅜	Nil. Nil.	1⅛ ⅞	Nil. Nil.	1⅜ 1⁹⁄₁₆	⅜ Nil.	1⅞ 1⅞	1⅜ 1⁵⁄₁₆

Tests with Falling Weight, 1120 lbs. falling 15 feet.

Date.	No. of Rails Tested.	Deflexion after one Blow.	Deflexion after two Blows.
March 8	6	1⅞ 1⅝ 2⁵⁄₁₆ 2⅛ 2⅝ 2⁵⁄₁₆	3⅜ 3¼ 3¼ 3⁷⁄₁₆ 4⅜ 3⅜

STATEMENT OF RAILS TENDERED, ACCEPTED, AND REJECTED.

Date.	Number of Rails Tendered.	Number of Rails Accepted.	Number of Rails rejected, and Nature of Defects.							
			Finally.				Pro tem.			Total.
			Head.	Web.	Flange.	Foot.	Crooked.	Cut Ends.		
March 17	893	861	6	...	6	4	10	6		32
" 18	2,406	2,338	8	...	10	11	20	18		68
" 19	2,754	2,669	3	...	20	16	21	25		85
" 20	3,069	3,005	7	...	9	6	32	10		64
" 21	2,002	1,925	7	...	11	10	33	16		77
Total	11,124	10,788	31	...	56	47	116	78		328

Tensile Tests of Steel Rails and Sleepers.

Date of Testing.	No.	Blow.	Description.	Original		Breaking Weight		Fractured		Difference between Original and Fractured Areas.		Elongation in 4 Inches.	
				Section.	Area.	On Original Area.	Per Square Inch.	Section.	Area.	Area.	Per Cent.	Actual.	Per Cent.
				diam.				diam.					
March 31 ..	1	..	Test-pieces turned out of head of I.S.R. 41¼ lb. flat-footed steel rail.	0·565	·25	8·70	34·80	0·43	·145	·105	42	$\frac{15}{16}$	23·4
”	2	..		0·565	·25	9·77	39·08	0·415	·135	·115	46	$\frac{13}{16}$	20·3
May 4 ..	1	..		0·565	·25	10·00	40·00	0·43	·145	·105	42	$\frac{15}{16}$	23·6
” ..	2	..		0·565	·25	9·59	38·36	0·43	·145	·105	42	1	25·0
												in 2 inches.	
April 23 ..	18	..	Test-pieces cut out of steel transverse sleepers.	$1 \times \frac{1}{4}$	·250	9·25	37·0	$\frac{3}{4} \times \frac{3}{16}$	·141	·109	43·6	$\frac{9}{16}$	28·1
” ..	19	..		$1 \times \frac{1}{4}$	·250	9·00	36·0	$\frac{13}{16} \times \frac{3}{16}$	·152	·098	39·2	$\frac{9}{16}$	28·1
” ..	20	..		$1 \times \frac{17}{64}$	·266	9·40	35·3	$\frac{3}{4} \times \frac{3}{16}$	·141	·125	47·0	$\frac{19}{32}$	29·7
April 24 ..	21	..		$1 \times \frac{7}{32}$	·219	8·50	38·8	$\frac{13}{16} \times \frac{5}{32}$	·127	·092	42·0	$\frac{17}{32}$	26·6
” ..	22	..		$1 \times \frac{1}{4}$	·250	8·05	32·2	$\frac{13}{16} \times \frac{1}{8}$	·102	·148	59·2	$\frac{17}{32}$	26·6
” ..	23	..		$1 \times \frac{3}{16}$	·188	7·85	41·7	$\frac{3}{4} \times \frac{1}{8}$	·094	·094	50·0	$\frac{1}{2}$	25·0

STANDARD TYPES OF PERMANENT WAY, INDIAN STATE RAILWAYS.
Quantities of Material in one Rail-length of Track.

Description of Permanent Ways.	Rails, 24 Feet.	Fish-plates, Pairs.	Fish-bolts.	Wood Sleepers.	*Bearing Plates.	Spikes.	C.I. Chairs with Pressed Wood Keys.	Steel Sleepers with Split Steel Keys or Wrought-iron Fastenings.	C.I. Pot Sleepers, with Tie-bars, Gibs, and Cotters, and Pressed Wood Keys.
5 ft. 8 in. gauge.									
75 lb. F.F. steel rails; wood sleepers, 10 ft. by 10 in. by 5 in.	2	2	12	8	16	36	..	..	..
75 lb. F.F. steel rails; stamped steel sleepers and steel keys.	2	2	12	..	..	..	..	8	..
75 lb. D.H. steel rails; wood sleepers, 10 ft. by 10 in. by 5 in.	2	2	12	8	..	32	16	..	..
75 lb. D.H. steel rails; cast iron pot sleepers.	2	2	12	..	..	..	..	..	8
Metre gauge.									
41¼ lb. F.F. steel rails; wood sleepers, 6 ft. by 8 in. by 5 in.	2	2	8	9	18	40	..	..	..
41¼ lb. F.F. steel rails; steel plate sleepers, with wrought iron rail fastenings.	2	2	8	..	..	..	..	9	..

* When good hard sleepers are used, bearing plates are only required for joint sleepers; on curves, however, the full complement may be allowed.

TABLE giving Cant in inches for different Curves and Maximum Speeds.

Radius.	5′ 6″ Gauge.						3′ 3⅜″ Gauge.						4′ 8½″ Gauge.					
	Maximum Speed in Miles per Hour.						Maximum Speed in Miles per Hour.						Maximum Speed in Miles per Hour.					
	10	20	25	30	40	50	10	15	20	25	30	40	20	25	30	40	50	60
400	1	..	..	..	..	..	$\frac{5}{8}$	$1\frac{1}{2}$	$2\frac{5}{8}$	$4\frac{1}{8}$	..	..	$3\frac{11}{32}$	..	..	..	..	..
600	$\frac{3}{4}$	$2\frac{15}{16}$	$4\frac{9}{16}$	$6\frac{5}{8}$	..	..	$\frac{7}{16}$	1	$1\frac{3}{4}$	$2\frac{3}{4}$	$3\frac{15}{16}$	..	$2\frac{11}{16}$	$4\frac{3}{16}$	$6\frac{1}{16}$	..	..	..
800	$\frac{9}{16}$	$2\frac{3}{16}$	$3\frac{7}{16}$	5	..	..	..	$\frac{3}{4}$	$1\frac{5}{16}$	$2\frac{1}{16}$	$2\frac{15}{16}$	$5\frac{1}{4}$	$1\frac{7}{8}$	$2\frac{15}{16}$	$4\frac{1}{4}$	..	..	..
1000	$\frac{7}{16}$	$1\frac{3}{4}$	$2\frac{3}{4}$	4	$7\frac{3}{32}$	..	..	$\frac{19}{32}$	$1\frac{1}{16}$	$1\frac{5}{8}$	$2\frac{3}{8}$	$4\frac{13}{16}$	$1\frac{1}{2}$	$2\frac{3}{8}$	$3\frac{3}{8}$	$6\frac{1}{32}$	..	..
1500	..	$1\frac{3}{16}$	$1\frac{13}{16}$	$2\frac{5}{16}$	$4\frac{11}{16}$	..	..	..	$\frac{11}{16}$	$1\frac{3}{32}$	$1\frac{9}{16}$	$2\frac{13}{32}$	1	$1\frac{9}{16}$	$2\frac{1}{4}$	4	$6\frac{9}{32}$	..
2000	..	$\frac{7}{8}$	$1\frac{3}{8}$	2	$3\frac{1}{2}$	$5\frac{1}{2}$	..	..	$\frac{1}{2}$	$\frac{13}{16}$	$1\frac{3}{16}$	$2\frac{3}{32}$	$\frac{3}{4}$	$1\frac{3}{16}$	$1\frac{11}{16}$	3	$4\frac{11}{16}$	$6\frac{25}{32}$
2500	..	$\frac{11}{16}$	$1\frac{1}{16}$	$1\frac{9}{16}$	$2\frac{3}{16}$	$4\frac{13}{32}$	..	..	..	$\frac{21}{32}$	$\frac{15}{16}$	$1\frac{11}{16}$	$\frac{5}{8}$	$\frac{15}{16}$	$1\frac{11}{32}$	$2\frac{3}{8}$	$3\frac{3}{4}$	$5\frac{7}{16}$
5000	..	..	$\frac{9}{16}$	$\frac{13}{16}$	$1\frac{13}{32}$	$2\frac{3}{16}$	..	..	..	..	$\frac{15}{32}$	$\frac{27}{32}$	..	$\frac{15}{32}$	$\frac{11}{16}$	$1\frac{3}{16}$	$1\frac{7}{8}$	$2\frac{23}{32}$

TABLE OF CHORDS (for different Speeds) to be stretched on the Curve, the versines being equal to Cant required.

Any radius	17	$34\frac{1}{2}$	43	$51\frac{1}{2}$	69	86	13	20	27	33	40	53	32	40	48	$63\frac{1}{2}$	$79\frac{1}{2}$	$95\frac{1}{2}$

POINTS AND CROSSINGS.

TABLE OF NUMBERS, ANGLES, RADII, CURVE-LEADS, SWITCH-LEADS, AND LEADS OF CROSSINGS.

Number. I.	Angle. a.			Radius. R.	Curve-lead. L′.	Switch-lead. l′.	Lead. L.	
Gauge = 5′ 6″. Clearance = 4¼″. Switches = 12′ and 15′.								
	°	′	″				′	″
12	4	45	49	1592·24	132·23	33·57	98	8
11	5	11	40	1339·24	121·25	30·79	90	5½
10	5	42	38	1108·24	110·27	28·01	82	3
9	6	20	25	899·24	99·31	25·23	74	1
8	7	7	30	712·24	88·34	22·45	65	10¾
7	8	7	48	547·24	77·39	19·68	57	8½
6	9	27	44	404·23	66·46	16·91	49	6½
5	11	18	36	283·22	55·54	14·16	41	4½
Gauge = 3′ 3⅜″. Clearance = 3⅝″. Switches = 9′ and 12′.								
12	As above			949·82	78·88	23·95	54	11¼
11				798·90	72·33	21·97	50	4¼
10				661·10	65·78	19·98	45	9½
9				536·42	59·24	18·00	41	3
8				424·87	52·70	16·02	36	8¼
7				326·45	46·17	14·04	32	1½
6				241·14	39·64	12·07	27	6¾
5				168·95	33·13	10·10	23	0¼
Gange = 4′ 8½″. Clearance = 4½″ Switches = 12′ and 15′.								
12	As above			1363·05	113·20	31·97	81	2¾
11				1146·06	103·80	29·32	74	5¾
10				948·72	94·40	26·67	67	8¾
9				769·80	85·01	24·03	60	11⅝
8				609·72	75·63	21·38	54	3
7				468·47	66·25	18·74	47	6
6				346·05	56·89	16·11	40	9¼
5				242·46	47·55	13·48	34	0¾

TURNOUTS AND CROSSOVERS.

Number. I.	Curve-lead. L′.	Lead. L.	Straight Intermediate Portion. S. D = 12.	D = 13.	D = 14.	D = 15.	D = 16.	Formulæ.
			Gauge = 5′ 6″.					
	′ ″	′ ″	′ ″	′ ″	′ ″	′ ″	′ ″	
12	132 2¾	98 8	11 9¼	23 9¼	35 9¼	47 9¼	59 9¼	S = 12 D − 132·229
11	121 3	90 5½	10 9	21 9	32 9	43 9	54 9	S = 11 D − 121·249
10	110 3¼	82 3	9 8¾	19 8¾	29 8¾	39 8¾	49 8¾	S = 10 D − 110·274
9	99 3¾	74 1	8 8¼	17 8¼	26 8¼	35 8¼	44 8¼	S = 9 D − 99·305
8	88 4	65 10¾	7 8	15 8	23 8	31 8	39 8	S = 8 D − 88·343
7	77 4¾	57 8½	6 7¼	13 7¼	20 7¼	27 7¼	34 7¼	S = 7 D − 77·391
			Gauge = 3′ 3⅜″.					
12	78 10½	54 11¼	65 1½	77 1½	89 1½	101 1½	113 1½	S = 12 D − 78·878
11	72 4	50 4¼	59 8	70 8	81 8	92 8	103 8	S = 11 D − 72·328
10	65 9¼	45 9½	54 2½	64 2½	74 2½	84 2½	94 2½	S = 10 D − 65·782
9	59 3	41 3	48 9¼	57 9¼	66 9¼	75 9¼	84 9¼	S = 9 D − 59·238
8	52 8½	36 8¼	43 3½	51 3½	59 3½	67 3½	75 3½	S = 8 D − 52·699
7	46 2	32 1½	37 10	44 10	51 10	58 10	65 10	S = 7 D − 46·166
			Gauge = 4′ 8½″.					
12	113 2½	81 2¾	30 9¾	42 9¾	54 9¾	66 9¾	78 9¾	S = 12 D − 113·195
11	103 9½	74 5¾	28 2½	39 2½	50 2½	61 2½	72 2½	S = 11 D − 103·796
10	94 4¾	67 8¾	25 7¼	35 7¼	45 7¼	55 7¼	65 7¼	S = 10 D − 94·401
9	85 0	60 11¾	23 0	32 0	41 0	50 0	59 0	S = 9 D − 85·010
8	75 7½	54 3	20 4½	28 4½	36 4½	44 4½	52 4½	S = 8 D − 75·627
7	66 3	47 6	17 9	24 9	31 9	38 9	45 9	S = 7 D − 66·251

INDIAN STATE RAILWAYS, 5' 6" GAUGE.

Points and Crossings.

Number. N.	Angle. a.			Radius. R.	Curve-lead. L'.		Switch-lead. l'.		Lead. L.		Remarks.
	°	′	″		′	″	′	″	′	″	
5·2671	10	45	0	313·48	58	6	14	11	43	7	Old standard crossings, still in use.
7·3479	7	45	0	602·14	81	3	20	8	60	7	
9·9310	5	45	0	1,098·11	109	6	27	10	81	8	
6	9	27	44	404·23	66	5	16	10	49	7	New standard crossings.
8½	6	42	35	802·99	93	10	23	10	70	0	
12	4	45	49	1,592·24	132	3	33	7	98	8	

Turnouts and Crossovers.

Crossing.	Straight Intermediate Portion (Figures 5 and 6). S.											
	D = 13⅓.		D = 14.		D = 15.		D = 16.		D = 18.		D = 20.	
° ′	′	″	′	″	′	″	′	″	′	″	′	″
10 45	12	8	15	3	20	7	25	10	36	4	46	11
7 45	18	0	21	8	29	0	36	4	51	1	65	9
5 45	24	7	29	6	39	5	49	4	69	3	89	1
1 in 6	14	7	17	7	23	7	29	7	41	7	53	7
1 in 8½	20	11	25	2	33	8	42	2	59	2	76	2
1 in 12	29	9	35	9	47	9	59	9	83	9	107	9
° ′	Curved Intermediate Portion (Figures 7 and 8).											
10 45	11	10	14	2	18	9	23	1	31	6	39	5
7 45	15	11	19	9	25	7	31	8	43	6	54	7
5 45	23	3	27	8	36	4	44	8	60	7	75	8
1 in 6	13	9	16	5	21	8	26	8	36	3	45	3
1 in 8½	19	9	23	7	30	11	38	1	51	8	64	7
1 in 12	28	2	33	6	43	11	54	0	73	4	91	6

INDIAN STATE RAILWAYS, 3' 3½" GAUGE.

Points and Crossings.

Number. 1.	Angle. a.			Radius. R.	Curve-lead. L'.		Switch-lead. l'.		Lead. L.		Remarks.
	°	′	″		′	″	′	″	′	″	
5·1446	11	0	0	178·50	34	1	10	5	23	8	Old standard crossings, still in use.
7·3479	7	45	0	359·19	48	5	14	8	33	9	
10·3854	5	30	0	712·65	68	4	20	9	47	7	
6¾	8	25	38	303·86	44	6	13	6	31	0	New standard crossings.
9½	6	0	32	597·14	62	6	19	0	43	6	
12	4	45	49	949·83	78	11	24	0	54	11	

Turnouts and Crossovers.

Crossing.	Straight Intermediate Portion (Figures 5 and 6.) S.					
	D = 13.	D = 14.	D = 15.	D = 16.	D = 18.	D = 20.
° ′	′ ″	′ ″	′ ″	′ ″	′ ″	′ ″
11 0	32 7	37 9	43 1	43 3	58 6	68 10
7 45	47 1	54 5	61 9	69 2	83 10	98 6
5 30	66 8	77 1	87 6	97 10	118 8	139 5
1 in 6¾	41 3	48 0	54 9	61 6	75 0	88 6
1 in 9½	61 0	70 6	80 0	89 6	108 6	127 6
1 in 12	77 1	89 1	101 1	113 1	137 1	161 1
° ′	Curved Intermediate Portion (Figures 7 and 8.)					
11 0	26 10	30 4	33 9	37 0	43 3	49 1
7 45	38 11	44 0	49 0	53 7	62 7	71 1
5 30	55 2	62 5	69 4	76 1	88 10	101 0
1 in 6¾	35 8	40 4	44 10	49 2	57 5	65 2
1 in 9½	50 6	57 1	63 5	69 7	81 3	92 4
1 in 12	63 10	72 2	80 2	88 0	102 10	116 10

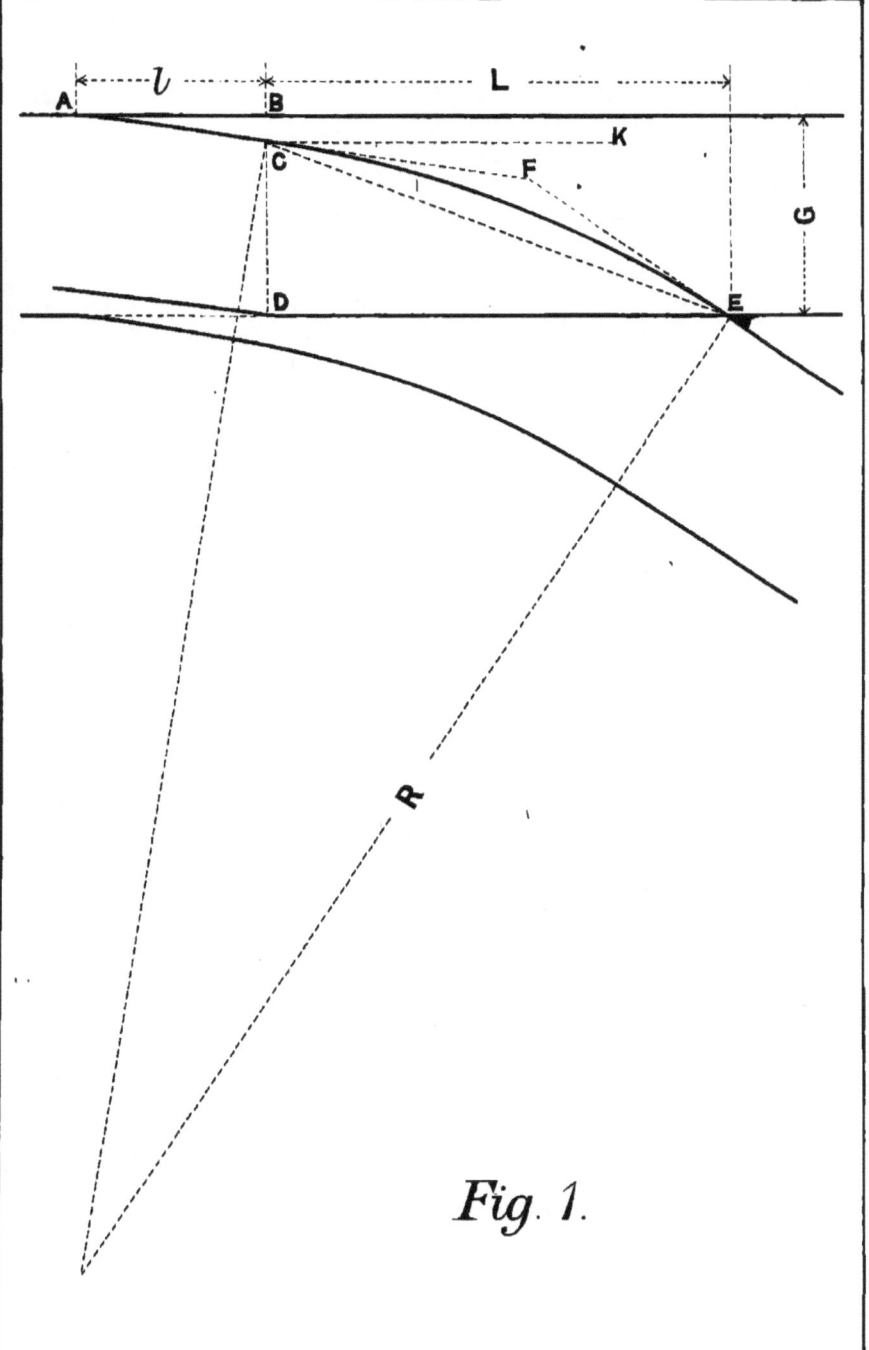

Fig. 1.

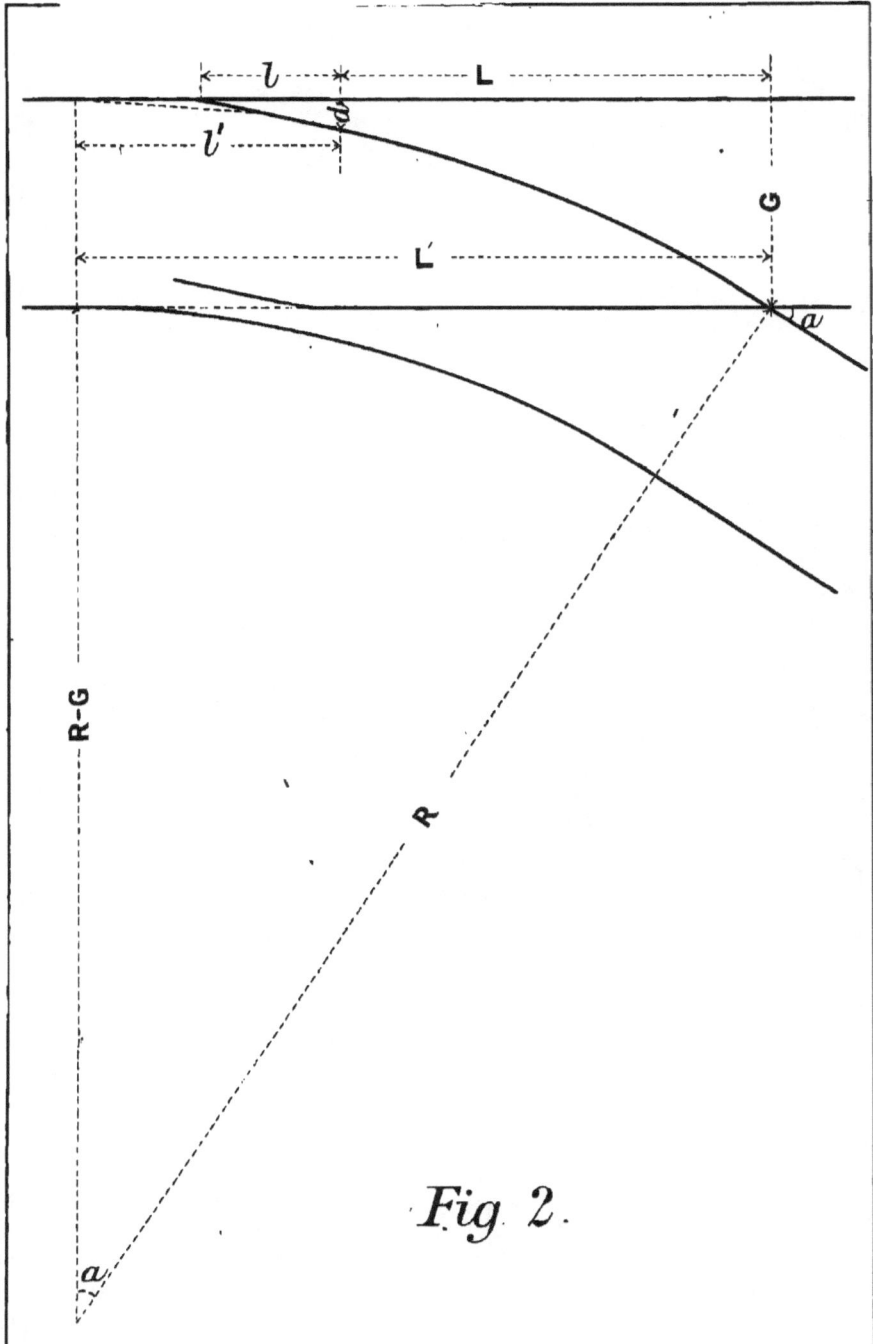

Fig 2.

PLATE. 4.

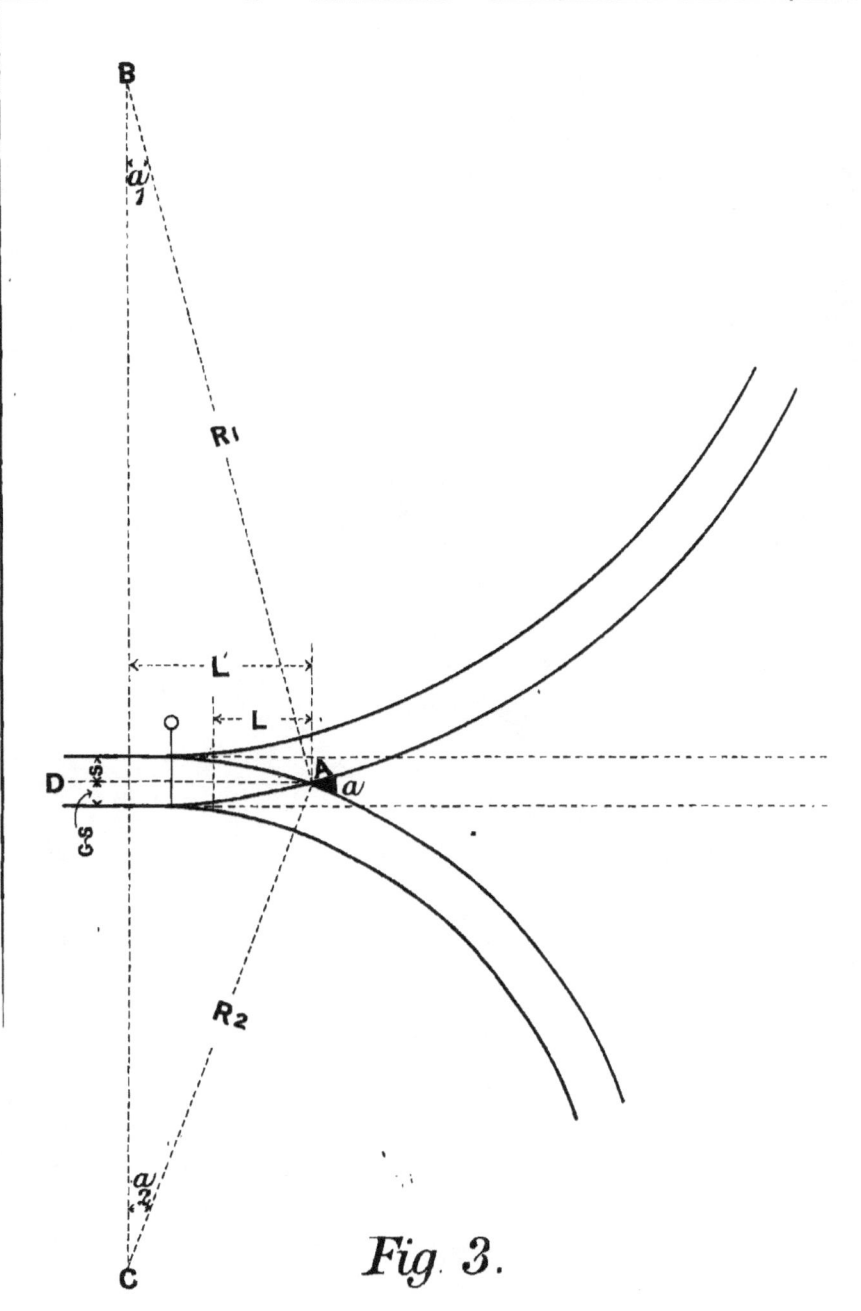

Fig. 3.

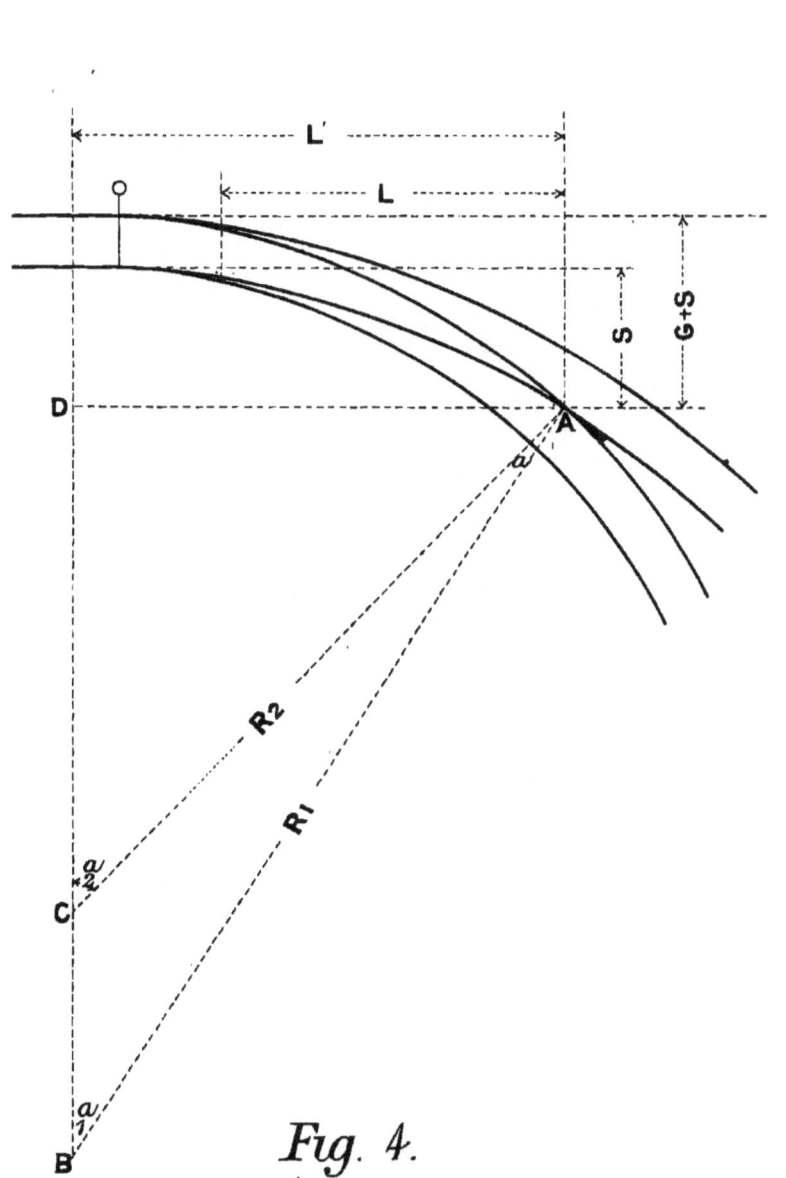

Fig. 4.

PLATE. 6.

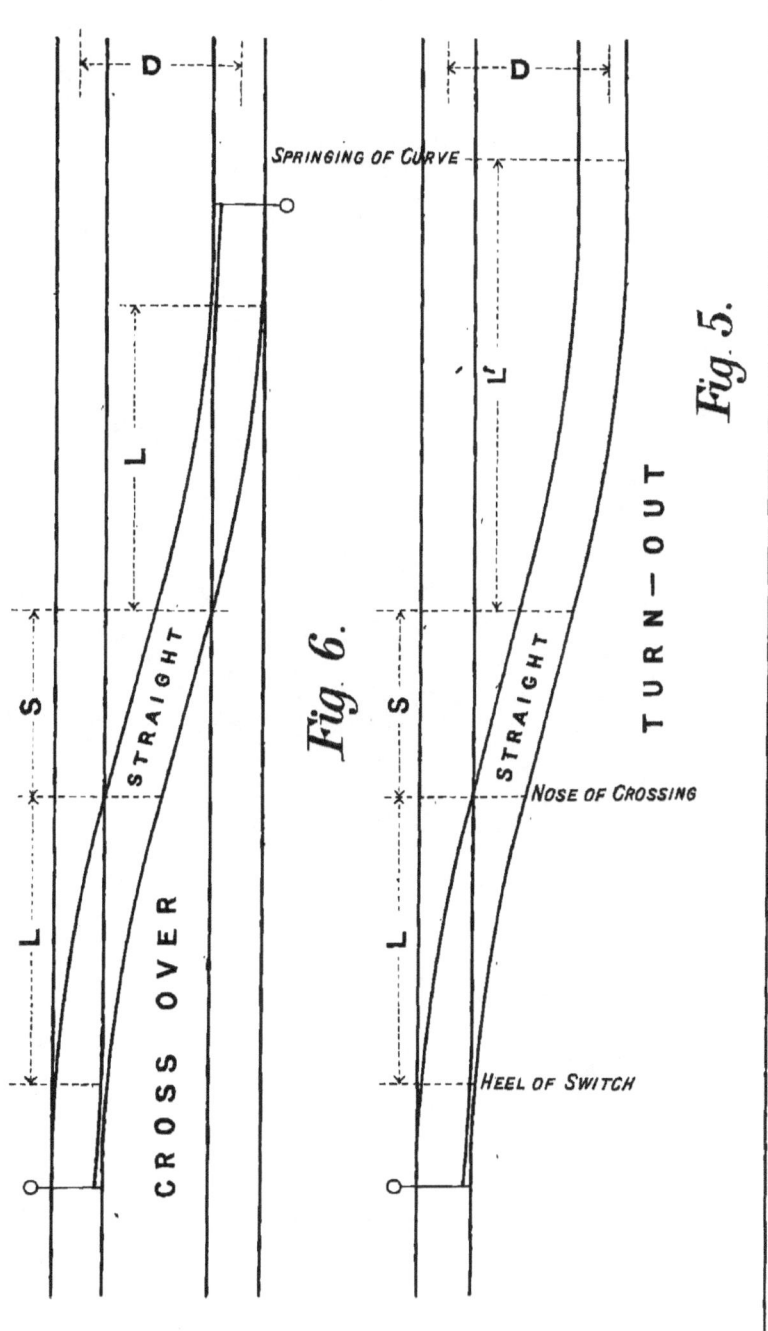

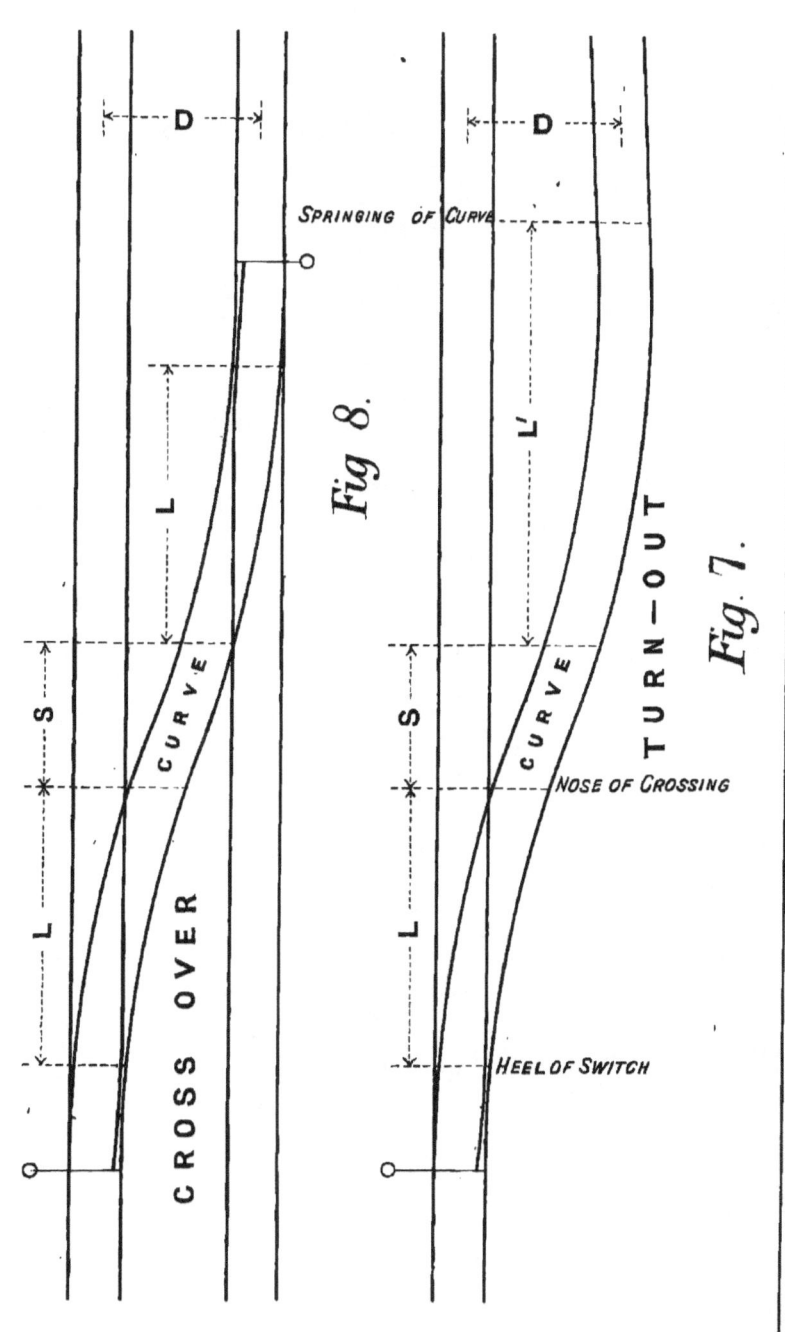

PLATE. 8.

Fig. 9.

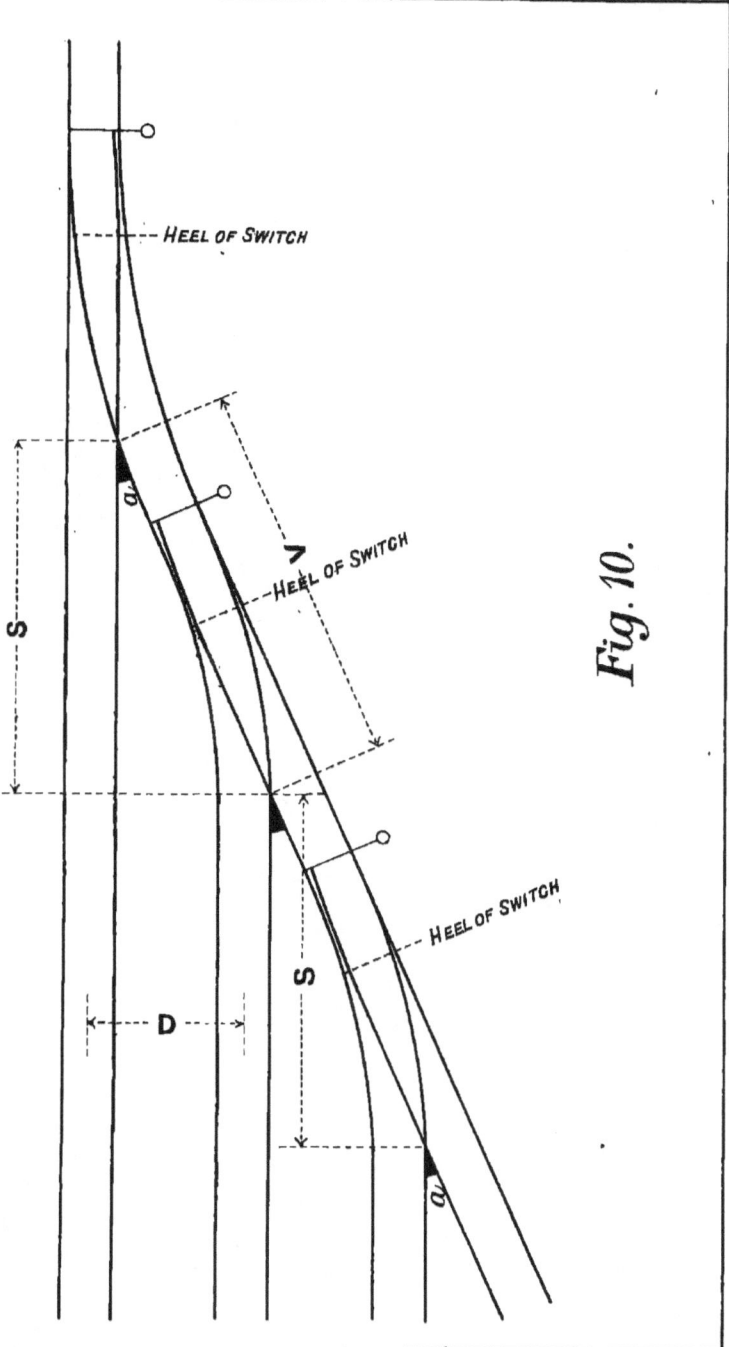

PLATE. 9.

Fig. 10.

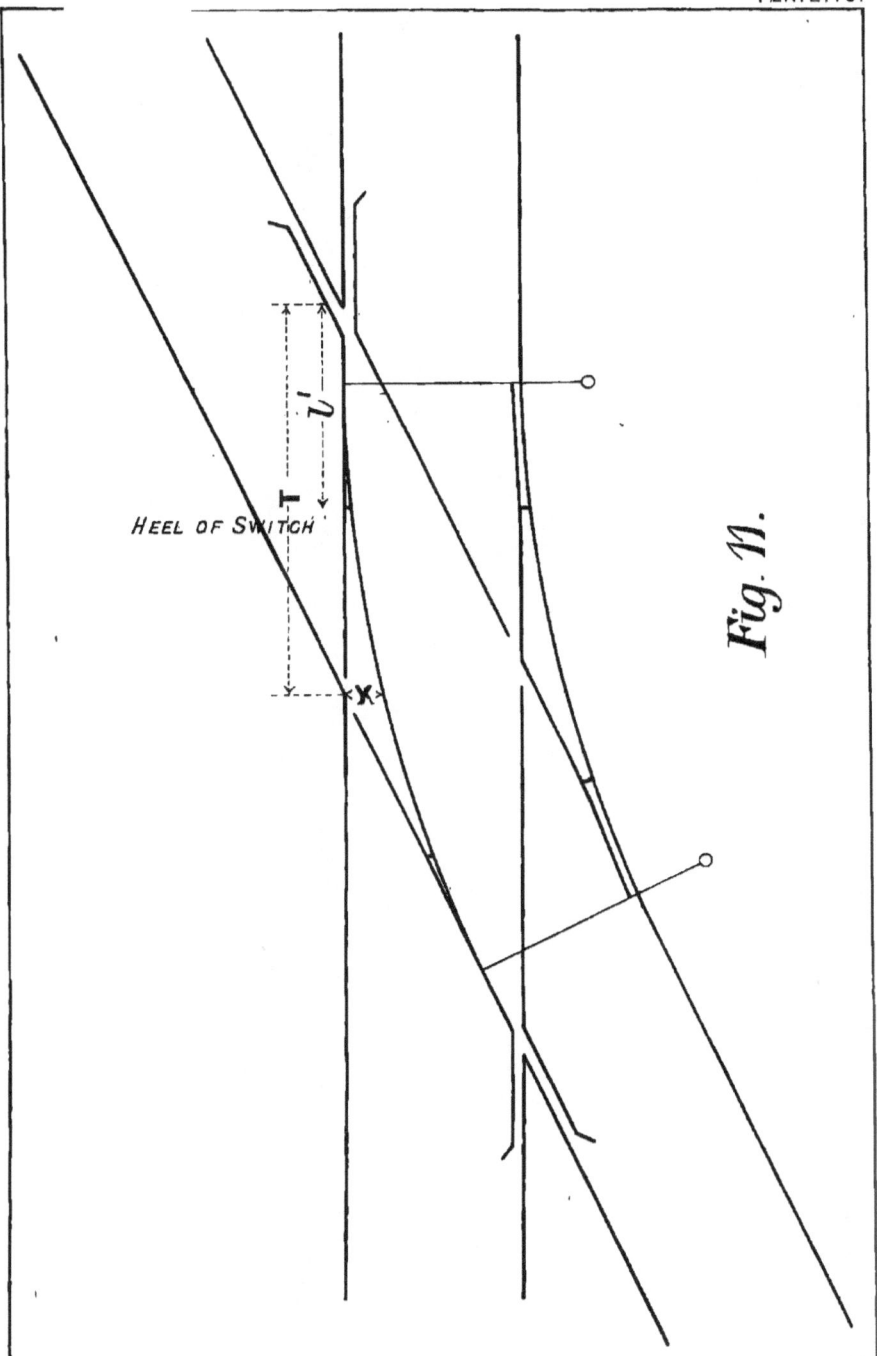

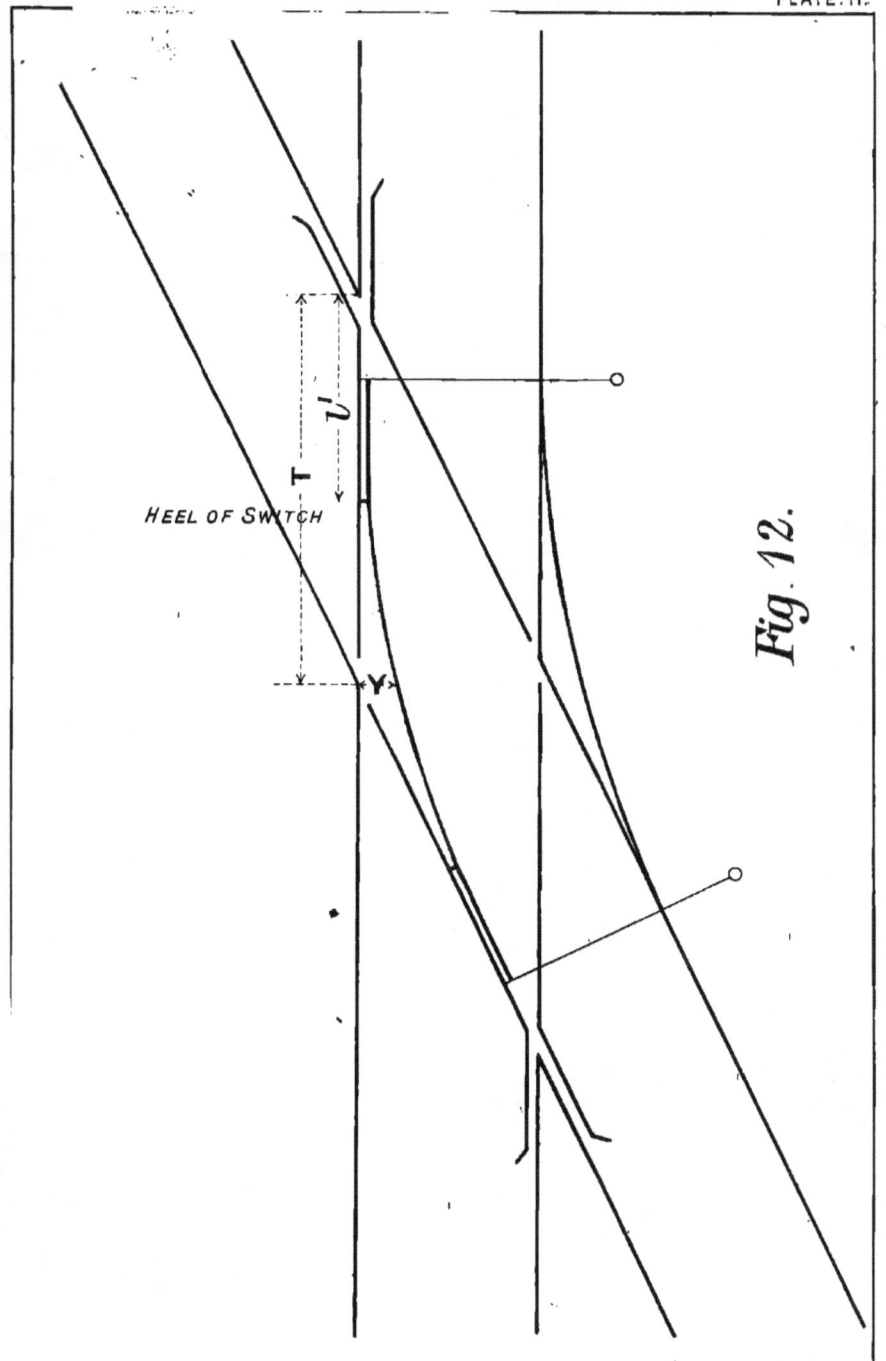

Fig. 12.

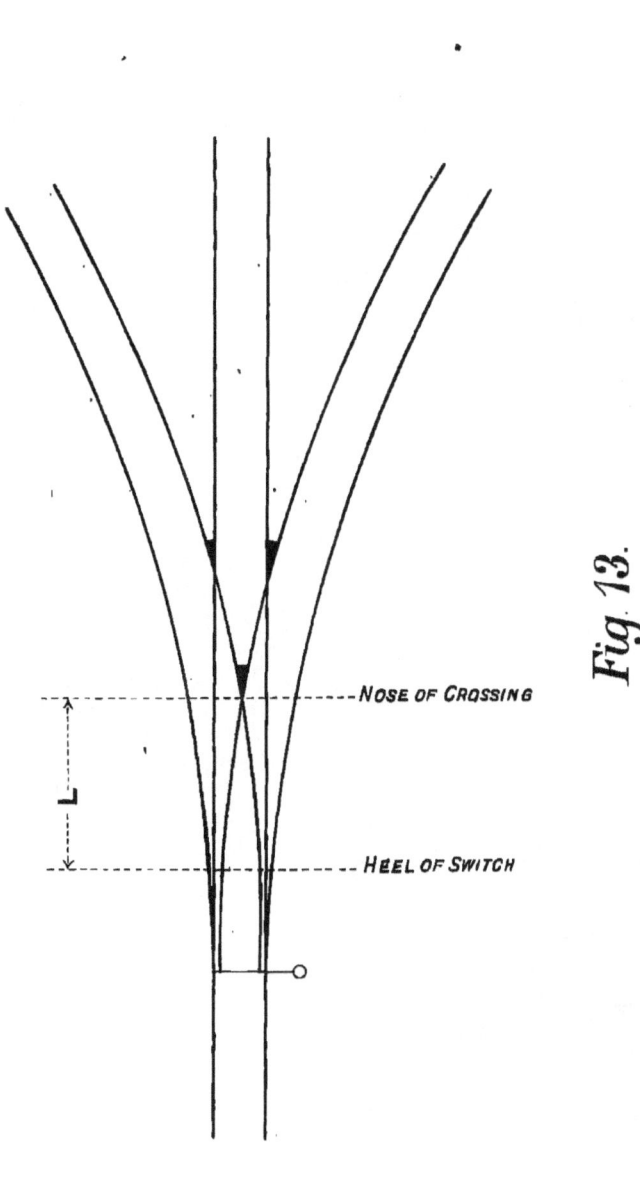

PLATE. 12.

Fig. 13.

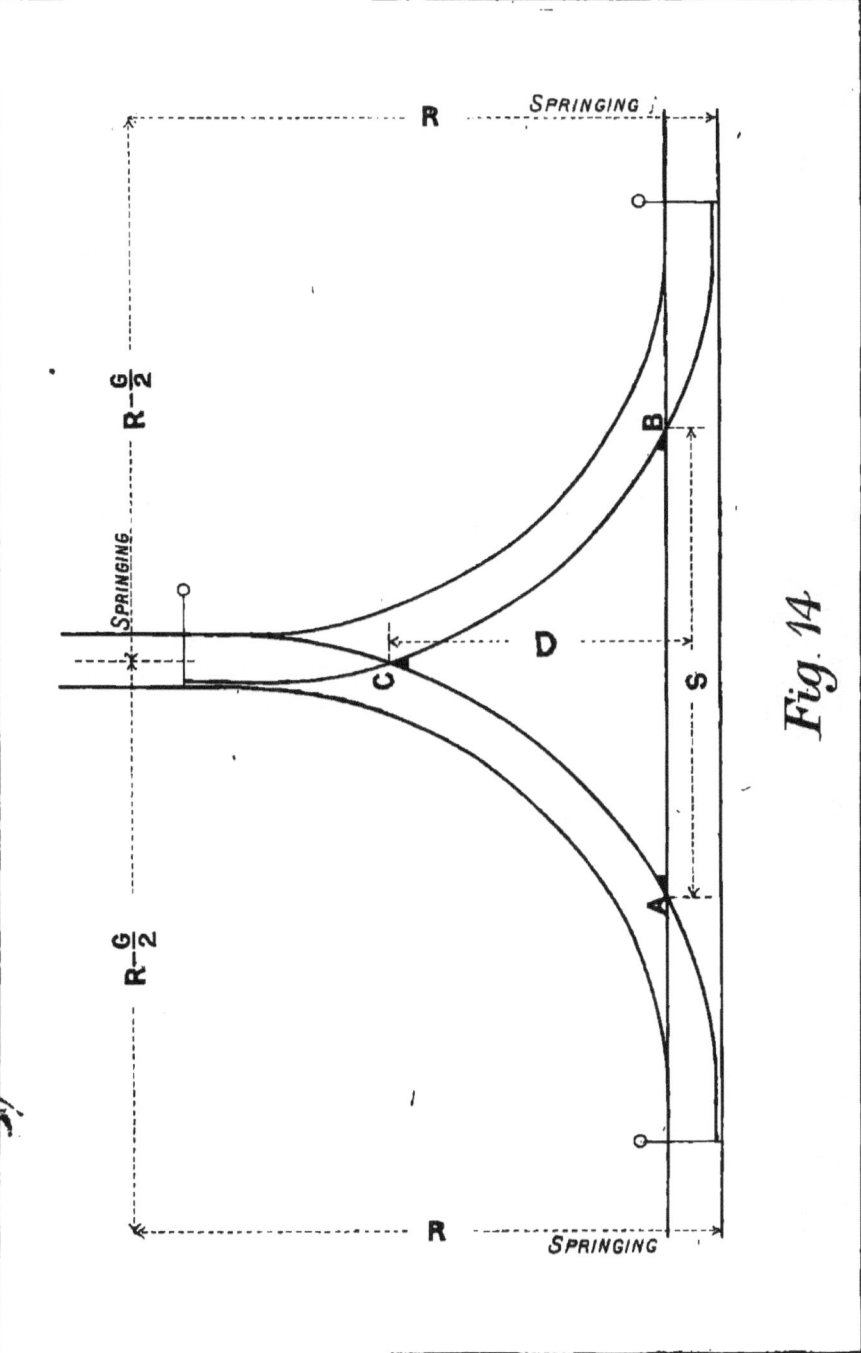

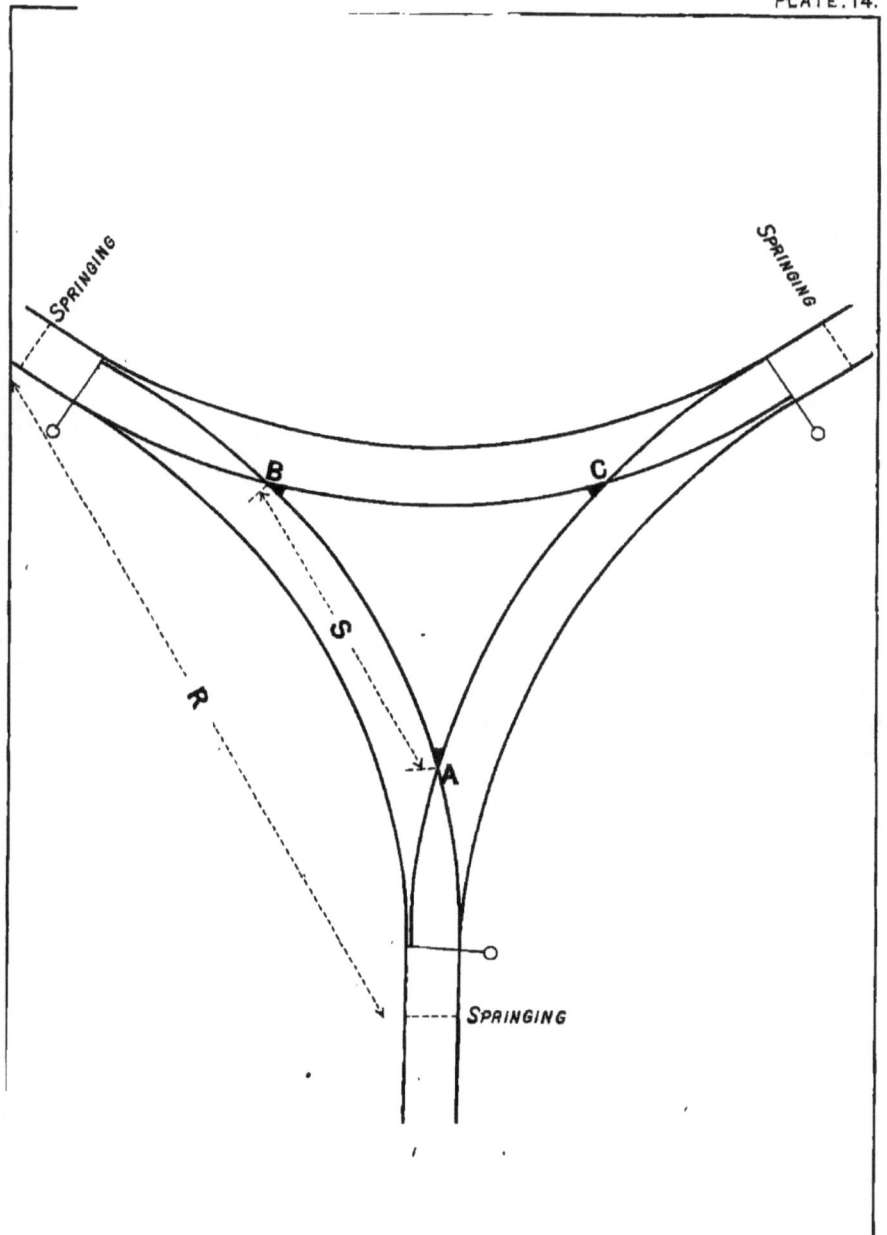

PLATE. 14.

Fig 15.

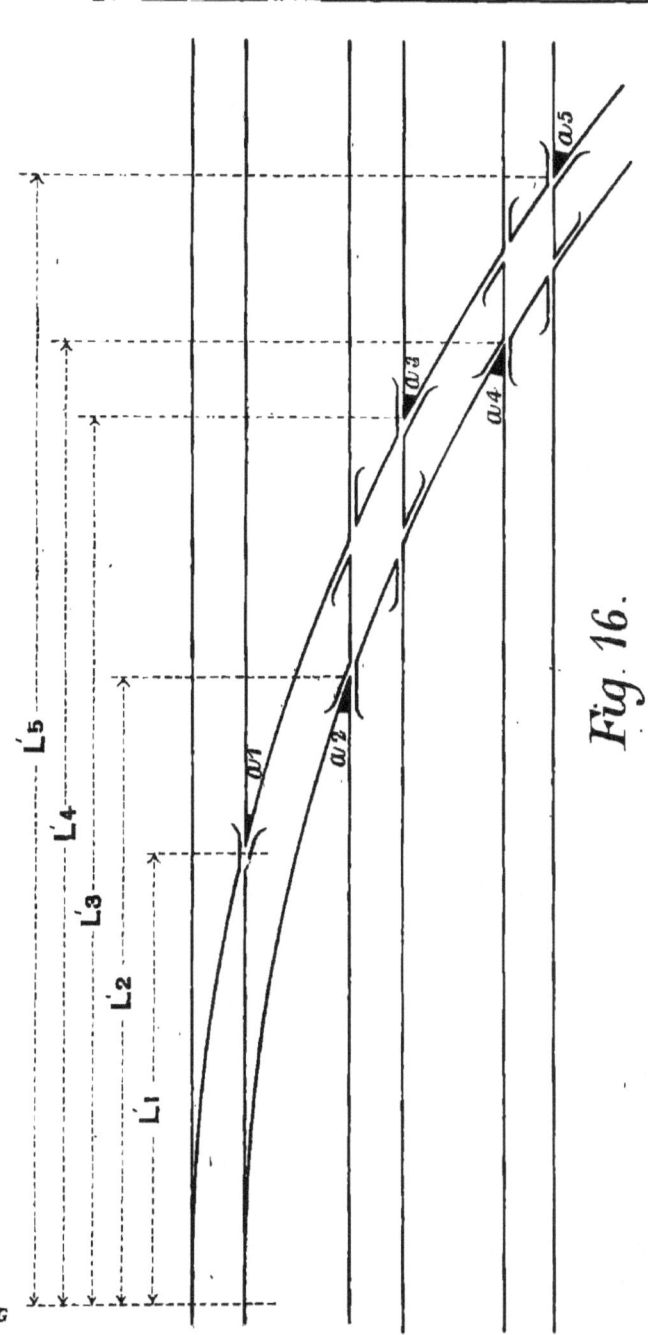

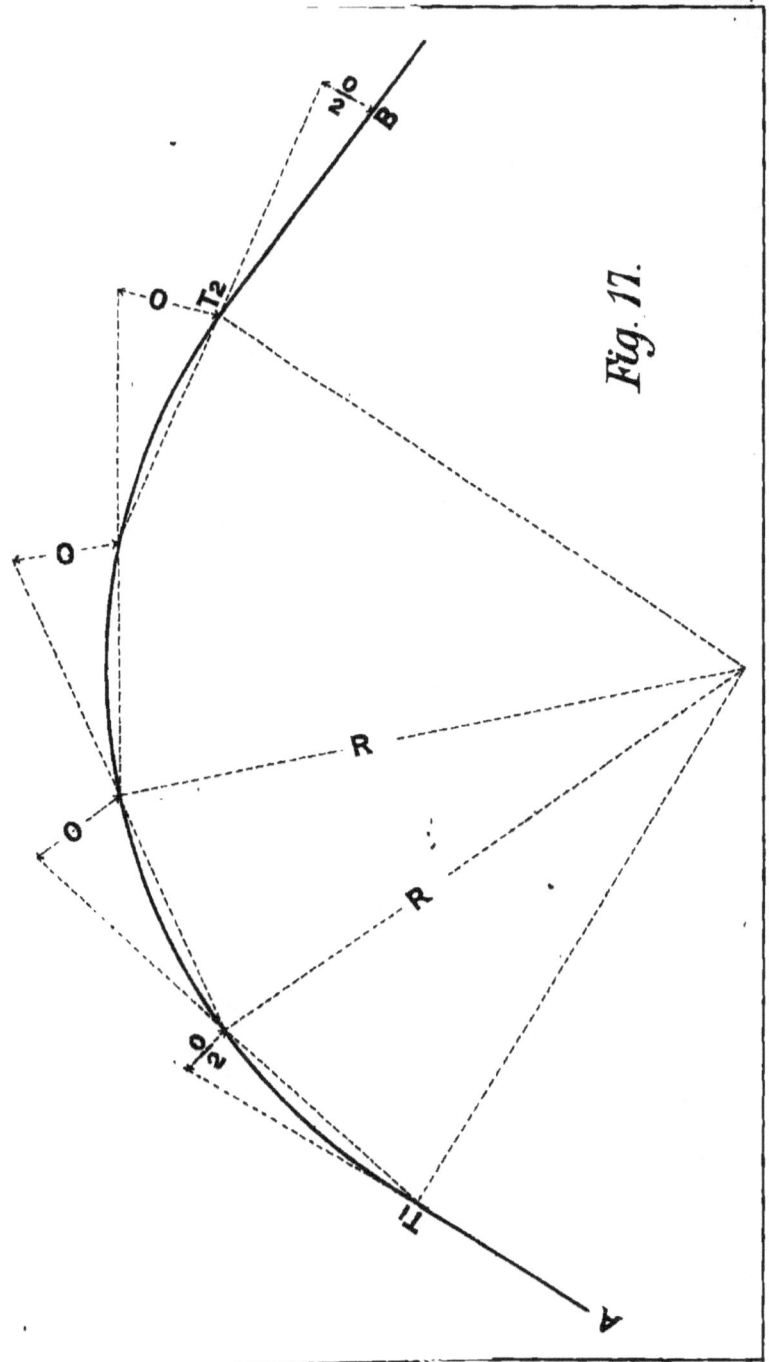

PLATE. 17.

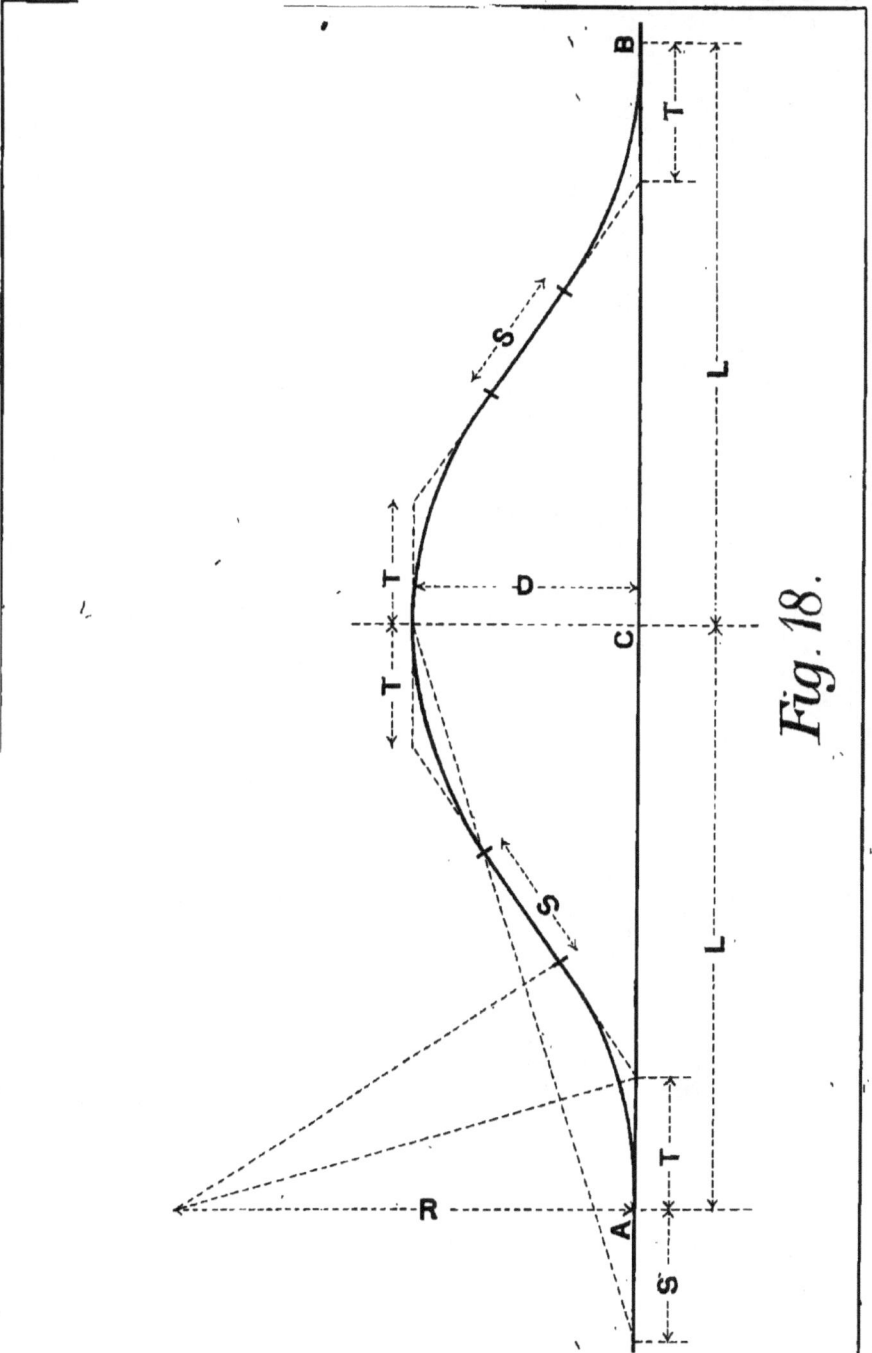

Fig. 18.

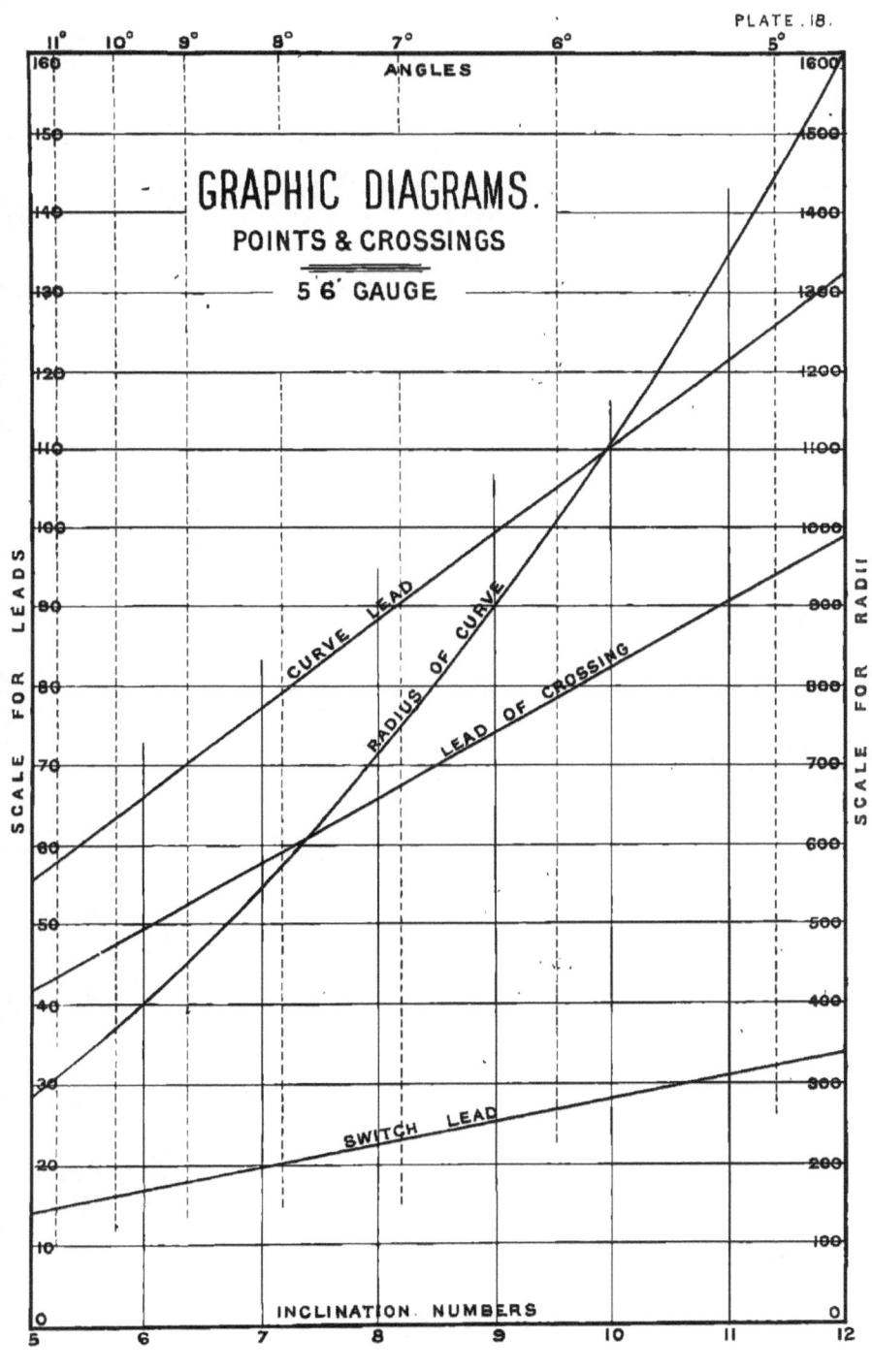

1890.

BOOKS RELATING

TO

APPLIED SCIENCE,

PUBLISHED BY

E. & F. N. SPON,

LONDON: 125, STRAND.

NEW YORK: 12, CORTLANDT STREET

The Engineers' Sketch-Book of Mechanical Movements, Devices, Appliances, Contrivances, Details employed in the Design and Construction of Machinery for every purpose. Collected from numerous Sources and from Actual Work. Classified and Arranged for Reference. Nearly 2000 *Illustrations*. By T. B. BARBER, Engineer. 8vo, cloth, 7s. 6d.

A Pocket-Book for Chemists, Chemical Manufacturers, Metallurgists, Dyers, Distillers, Brewers, Sugar Refiners, Photographers, Students, etc., etc. By THOMAS BAYLEY, Assoc. R.C. Sc. Ireland, Analytical and Consulting Chemist and Assayer. Fourth edition, with additions, 437 pp., royal 32mo, roan, gilt edges, 5s.

SYNOPSIS OF CONTENTS:

Atomic Weights and Factors—Useful Data—Chemical Calculations—Rules for Indirect Analysis—Weights and Measures—Thermometers and Barometers—Chemical Physics—Boiling Points, etc.—Solubility of Substances—Methods of Obtaining Specific Gravity—Conversion of Hydrometers—Strength of Solutions by Specific Gravity—Analysis—Gas Analysis—Water Analysis—Qualitative Analysis and Reactions—Volumetric Analysis—Manipulation—Mineralogy—Assaying—Alcohol—Beer—Sugar—Miscellaneous Technological matter relating to Potash, Soda, Sulphuric Acid, Chlorine, Tar Products, Petroleum, Milk, Tallow, Photography, Prices, Wages, Appendix, etc., etc.

The Mechanician: A Treatise on the Construction and Manipulation of Tools, for the use and instruction of Young Engineers and Scientific Amateurs, comprising the Arts of Blacksmithing and Forging; the Construction and Manufacture of Hand Tools, and the various Methods of Using and Grinding them; description of Hand and Machine Processes; Turning and Screw Cutting. By CAMERON KNIGHT, Engineer. Containing 1147 *illustrations*, and 397 pages of letter-press. Fourth edition, 4to, cloth, 18s.

B

CATALOGUE OF SCIENTIFIC BOOKS

Just Published, in Demy 8vo, cloth, containing 975 pages and 250 Illustrations, price 7s. 6d.

SPONS' HOUSEHOLD MANUAL:
A Treasury of Domestic Receipts and Guide for Home Management.

PRINCIPAL CONTENTS.

Hints for selecting a good House, pointing out the essential requirements for a good house as to the Site, Soil, Trees, Aspect, Construction, and General Arrangement; with instructions for Reducing Echoes, Waterproofing Damp Walls, Curing Damp Cellars.

Sanitation.—What should constitute a good Sanitary Arrangement; Examples (with Illustrations) of Well- and Ill-drained Houses; How to Test Drains; Ventilating Pipes, etc.

Water Supply.—Care of Cisterns; Sources of Supply; Pipes; Pumps; Purification and Filtration of Water.

Ventilation and Warming.—Methods of Ventilating without causing cold draughts, by various means; Principles of Warming; Health Questions; Combustion; Open Grates; Open Stoves; Fuel Economisers; Varieties of Grates; Close-Fire Stoves; Hot-air Furnaces; Gas Heating; Oil Stoves; Steam Heating; Chemical Heaters; Management of Flues; and Cure of Smoky Chimneys.

Lighting.—The best methods of Lighting; Candles, Oil Lamps, Gas, Incandescent Gas, Electric Light; How to test Gas Pipes; Management of Gas.

Furniture and Decoration.—Hints on the Selection of Furniture; on the most approved methods of Modern Decoration; on the best methods of arranging Bells and Calls; How to Construct an Electric Bell.

Thieves and Fire.—Precautions against Thieves and Fire; Methods of Detection; Domestic Fire Escapes; Fireproofing Clothes, etc.

The Larder.—Keeping Food fresh for a limited time; Storing Food without change, such as Fruits, Vegetables, Eggs, Honey, etc.

Curing Foods for lengthened Preservation, as Smoking, Salting, Canning, Potting, Pickling, Bottling Fruits, etc.; Jams, Jellies, Marmalade, etc.

The Dairy.—The Building and Fitting of Dairies in the most approved modern style; Butter-making; Cheesemaking and Curing.

The Cellar.—Building and Fitting; Cleaning Casks and Bottles; Corks and Corking; Aërated Drinks; Syrups for Drinks; Beers; Bitters; Cordials and Liqueurs; Wines; Miscellaneous Drinks.

The Pantry.—Bread-making; Ovens and Pyrometers; Yeast; German Yeast; Biscuits; Cakes; Fancy Breads; Buns.

The Kitchen.—On Fitting Kitchens; a description of the best Cooking Ranges, close and open; the Management and Care of Hot Plates, Baking Ovens, Dampers, Flues, and Chimneys; Cooking by Gas; Cooking by Oil; the Arts of Roasting, Grilling, Boiling, Stewing, Braising, Frying.

Receipts for Dishes—Soups, Fish, Meat, Game, Poultry, Vegetables, Salads, Puddings, Pastry, Confectionery, Ices, etc., etc.; Foreign Dishes.

The Housewife's Room.—Testing Air, Water, and Foods; Cleaning and Renovating; Destroying Vermin.

Housekeeping, Marketing.

The Dining-Room.—Dietetics; Laying and Waiting at Table; Carving; Dinners, Breakfasts, Luncheons, Teas, Suppers, etc.

The Drawing-Room.—Etiquette; Dancing; Amateur Theatricals; Tricks and Illusions; Games (indoor).

The Bedroom and Dressing-Room; Sleep; the Toilet; Dress; Buying Clothes; Outfits; Fancy Dress.

The Nursery.—The Room; Clothing; Washing; Exercise; Sleep; Feeding; Teething; Illness; Home Training.

The Sick-Room.—The Room; the Nurse; the Bed; Sick Room Accessories; Feeding Patients; Invalid Dishes and Drinks; Administering Physic; Domestic Remedies; Accidents and Emergencies; Bandaging; Burns; Carrying Injured Persons; Wounds; Drowning; Fits; Frost-bites; Poisons and Antidotes; Sunstroke; Common Complaints; Disinfection, etc.

The Bath-Room.—Bathing in General; Management of Hot-Water System.
The Laundry.—Small Domestic Washing Machines, and methods of getting up linen; Fitting up and Working a Steam Laundry.
The School-Room.—The Room and its Fittings; Teaching, etc.
The Playground.—Air and Exercise; Training; Outdoor Games and Sports.
The Workroom.—Darning, Patching, and Mending Garments.
The Library.—Care of Books.
The Garden.—Calendar of Operations for Lawn, Flower Garden, and Kitchen Garden.
The Farmyard.—Management of the Horse, Cow, Pig, Poultry, Bees, etc., etc.
Small Motors.—A description of the various small Engines useful for domestic purposes, from 1 man to 1 horse power, worked by various methods, such as Electric Engines, Gas Engines, Petroleum Engines, Steam Engines, Condensing Engines, Water Power, Wind Power, and the various methods of working and managing them.
Household Law.—The Law relating to Landlords and Tenants, Lodgers, Servants, Parochial Authorities, Juries, Insurance, Nuisance, etc.

On Designing Belt Gearing. By E. J. COWLING WELCH, Mem. Inst. Mech. Engineers, Author of 'Designing Valve Gearing.' Fcap. 8vo, sewed, 6d.

A Handbook of Formulæ, Tables, and Memoranda, for Architectural Surveyors and others engaged in Building. By J. T. HURST, C.E. Fourteenth edition, royal 32mo, roan, 5s.

"It is no disparagement to the many excellent publications we refer to, to say that in our opinion this little pocket-book of Hurst's is the very best of them all, without any exception. It would be useless to attempt a recapitulation of the contents, for it appears to contain almost *everything* that anyone connected with building could require, and, best of all, made up in a compact form for carrying in the pocket, measuring only 5 in. by 3 in., and about ¼ in. thick, in a limp cover. We congratulate the author on the success of his laborious and practically compiled little book, which has received unqualified and deserved praise from every professional person to whom we have shown it."—*The Dublin Builder.*

Tabulated Weights of Angle, Tee, Bulb, Round, Square, and Flat Iron and Steel, and other information for the use of Naval Architects and Shipbuilders. By C. H. JORDAN, M.I.N.A. Fourth edition, 32mo, cloth, 2s. 6d.

A Complete Set of Contract Documents for a Country Lodge, comprising Drawings, Specifications, Dimensions (for quantities), Abstracts, Bill of Quantities, Form of Tender and Contract, with Notes by J. LEANING, printed in facsimile of the original documents, on single sheets fcap., in paper case, 10s.

A Practical Treatise on Heat, as applied to the Useful Arts; for the Use of Engineers, Architects, &c. By THOMAS BOX. *With 14 plates.* Sixth edition, crown 8vo, cloth, 12s. 6d.

A Descriptive Treatise on Mathematical Drawing Instruments: their construction, uses, qualities, selection, preservation, and suggestions for improvements, with hints upon Drawing and Colouring. By W. F. STANLEY, M.R.I. Sixth edition, *with numerous illustrations,* crown 8vo, cloth, 5s.

B 2

Quantity Surveying. By J. LEANING. With 42 illustrations. Second edition, revised, crown 8vo, cloth, 9s.

CONTENTS:

A complete Explanation of the London Practice.
General Instructions.
Order of Taking Off.
Modes of Measurement of the various Trades.
Use and Waste.
Ventilation and Warming.
Credits, with various Examples of Treatment.
Abbreviations.
Squaring the Dimensions.
Abstracting, with Examples in illustration of each Trade.
Billing.
Examples of Preambles to each Trade.
Form for a Bill of Quantities.
   Do.   Bill of Credits.
   Do.   Bill for Alternative Estimate.
Restorations and Repairs, and Form of Bill.
Variations before Acceptance of Tender.
Errors in a Builder's Estimate.

Schedule of Prices.
Form of Schedule of Prices.
Analysis of Schedule of Prices.
Adjustment of Accounts.
Form of a Bill of Variations.
Remarks on Specifications.
Prices and Valuation of Work, with Examples and Remarks upon each Trade.
The Law as it affects Quantity Surveyors, with Law Reports.
Taking Off after the Old Method.
Northern Practice.
The General Statement of the Methods recommended by the Manchester Society of Architects for taking Quantities.
Examples of Collections.
Examples of "Taking Off" in each Trade.
Remarks on the Past and Present Methods of Estimating.

Spons' Architects' and Builders' Price Book, with useful Memoranda. Edited by W. YOUNG, Architect. Crown 8vo, cloth, red edges, 3s. 6d. *Published annually.* Seventeenth edition. *Now ready.*

Long-Span Railway Bridges, comprising Investigations of the Comparative Theoretical and Practical Advantages of the various adopted or proposed Type Systems of Construction, with numerous Formulæ and Tables giving the weight of Iron or Steel required in Bridges from 300 feet to the limiting Spans; to which are added similar Investigations and Tables relating to Short-span Railway Bridges. Second and revised edition. By B. BAKER, Assoc. Inst. C.E. *Plates,* crown 8vo, cloth, 5s.

Elementary Theory and Calculation of Iron Bridges and Roofs. By AUGUST RITTER, Ph.D., Professor at the Polytechnic School at Aix-la-Chapelle. Translated from the third German edition, by H. R. SANKEY, Capt. R.E. With 500 *illustrations,* 8vo, cloth, 15s.

The Elementary Principles of Carpentry. By THOMAS TREDGOLD. Revised from the original edition, and partly re-written, by JOHN THOMAS HURST. Contained in 517 pages of letter-press, and *illustrated with 48 plates and 150 wood engravings.* Sixth edition, reprinted from the third, crown 8vo, cloth, 12s. 6d.

Section I. On the Equality and Distribution of Forces—Section II. Resistance of Timber—Section III. Construction of Floors—Section IV. Construction of Roofs—Section V. Construction of Domes and Cupolas—Section VI. Construction of Partitions—Section VII. Scaffolds, Staging, and Gantries—Section VIII. Construction of Centres for Bridges—Section IX. Coffer-dams, Shoring, and Strutting—Section X. Wooden Bridges and Viaducts—Section XI. Joints, Straps, and other Fastenings—Section XII. Timber.

The Builder's Clerk : a Guide to the Management of a Builder's Business. By THOMAS BALES. Fcap. 8vo, cloth, 1s. 6d.

Practical Gold-Mining: a Comprehensive Treatise on the Origin and Occurrence of Gold-bearing Gravels, Rocks and Ores, and the methods by which the Gold is extracted. By C. G. WARNFORD LOCK, co-Author of 'Gold; its Occurrence and Extraction.' *With 8 plates and* 275 *engravings in the text,* royal 8vo, cloth, 2*l.* 2*s.*

Hot Water Supply: A Practical Treatise upon the Fitting of Circulating Apparatus in connection with Kitchen Range and other Boilers, to supply Hot Water for Domestic and General Purposes. With a Chapter upon Estimating. *Fully illustrated,* crown 8vo, cloth, 3*s.*

Hot Water Apparatus: An Elementary Guide for the Fitting and Fixing of Boilers and Apparatus for the Circulation of Hot Water for Heating and for Domestic Supply, and containing a Chapter upon Boilers and Fittings for Steam Cooking. 32 *illustrations,* fcap. 8vo, cloth, 1*s.* 6*d.*

The Use and Misuse, and the Proper and Improper Fixing of a Cooking Range. Illustrated, fcap. 8vo, sewed, 6*d.*

Iron Roofs: Examples of Design, Description. *Illustrated with* 64 *Working Drawings of Executed Roofs.* By ARTHUR T. WALMISLEY, Assoc. Mem. Inst. C.E. Second edition, revised, imp. 4to, half-morocco, 3*l.* 3*s.*

A History of Electric Telegraphy, to the Year 1837. Chiefly compiled from Original Sources, and hitherto Unpublished Documents, by J. J. FAHIE, Mem. Soc. of Tel. Engineers, and of the International Society of Electricians, Paris. Crown 8vo, cloth, 9*s.*

Spons' Information for Colonial Engineers. Edited by J. T. HURST. Demy 8vo, sewed.

No. 1, Ceylon. By ABRAHAM DEANE, C.E. 2*s.* 6*d.*

CONTENTS:

Introductory Remarks—Natural Productions—Architecture and Engineering—Topography, Trade, and Natural History—Principal Stations—Weights and Measures, etc., etc.

No. 2. Southern Africa, including the Cape Colony, Natal, and the Dutch Republics. By HENRY HALL, F.R.G.S., F.R.C.I. With Map. 3*s.* 6*d.*

CONTENTS:

General Description of South Africa—Physical Geography with reference to Engineering Operations—Notes on Labour and Material in Cape Colony—Geological Notes on Rock Formation in South Africa—Engineering Instruments for Use in South Africa—Principal Public Works in Cape Colony: Railways, Mountain Roads and Passes, Harbour Works, Bridges, Gas Works, Irrigation and Water Supply, Lighthouses, Drainage and Sanitary Engineering, Public Buildings, Mines—Table of Woods in South Africa—Animals used for Draught Purposes—Statistical Notes—Table of Distances—Rates of Carriage, etc.

No. 3. India. By F. C. DANVERS, Assoc. Inst. C.E. With Map. 4*s.* 6*d.*

CONTENTS:

Physical Geography of India—Building Materials—Roads—Railways—Bridges—Irrigation—River Works—Harbours—Lighthouse Buildings—Native Labour—The Principal Trees of India—Money—Weights and Measures—Glossary of Indian Terms, etc.

Our Factories, Workshops, and Warehouses: their Sanitary and Fire-Resisting Arrangements. By B. H. THWAITE, Assoc. Mem. Inst. C.E. *With 183 wood engravings*, crown 8vo, cloth, 9s.

A Practical Treatise on Coal Mining. By GEORGE G. ANDRÉ, F.G.S., Assoc. Inst. C.E., Member of the Society of Engineers. *With 82 lithographic plates.* 2 vols., royal 4to, cloth, 3l. 12s.

A Practical Treatise on Casting and Founding, including descriptions of the modern machinery employed in the art. By N. E. SPRETSON, Engineer. Fifth edition, with 82 *plates* drawn to scale, 412 pp., demy 8vo, cloth, 18s.

The Depreciation of Factories and their Valuation. By EWING MATHESON, M. Inst. C.E. 8vo, cloth, 6s.

A Handbook of Electrical Testing. By H. R. KEMPE, M.S.T.E. Fourth edition, revised and enlarged, crown 8vo, cloth, 16s.

The Clerk of Works: a Vade-Mecum for all engaged in the Superintendence of Building Operations. By G. G. HOSKINS, F.R.I.B.A. Third edition, fcap. 8vo, cloth, 1s. 6d.

American Foundry Practice: Treating of Loam, Dry Sand, and Green Sand Moulding, and containing a Practical Treatise upon the Management of Cupolas, and the Melting of Iron. By T. D. WEST, Practical Iron Moulder and Foundry Foreman. Second edition, *with numerous illustrations*, crown 8vo, cloth, 10s. 6d.

The Maintenance of Macadamised Roads. By T. CODRINGTON, M.I.C.E, F.G.S., General Superintendent of County Roads for South Wales. 8vo, cloth, 6s.

Hydraulic Steam and Hand Power Lifting and Pressing Machinery. By FREDERICK COLYER, M. Inst. C.E., M. Inst. M.E. *With 73 plates*, 8vo, cloth, 18s.

Pumps and Pumping Machinery. By F. COLYER, M.I.C.E., M.I.M.E. *With 23 folding plates*, 8vo, cloth, 12s. 6d.

Pumps and Pumping Machinery. By F. COLYER. Second Part. *With 11 large plates*, 8vo, cloth, 12s. 6d.

A Treatise on the Origin, Progress, Prevention, and Cure of Dry Rot in Timber; with Remarks on the Means of Preserving Wood from Destruction by Sea-Worms, Beetles, Ants, etc. By THOMAS ALLEN BRITTON, late Surveyor to the Metropolitan Board of Works, etc., etc. *With 10 plates*, crown 8vo, cloth, 7s. 6d.

Gas Works: their Arrangement, Construction, Plant, and Machinery. By F. COLYER, M. Inst. C.E. *With 31 folding plates,* 8vo, cloth, 12s. 6d.

The Municipal and Sanitary Engineer's Handbook. By H. PERCY BOULNOIS, Mem. Inst. C.E., Borough Engineer, Portsmouth. *With numerous illustrations,* demy 8vo, cloth, 12s. 6d.

CONTENTS:

The Appointment and Duties of the Town Surveyor—Traffic—Macadamised Roadways—Steam Rolling—Road Metal and Breaking—Pitched Pavements—Asphalte—Wood Pavements—Footpaths—Kerbs and Gutters—Street Naming and Numbering—Street Lighting—Sewerage—Ventilation of Sewers—Disposal of Sewage—House Drainage—Disinfection—Gas and Water Companies, etc., Breaking up Streets—Improvement of Private Streets—Borrowing Powers—Artizans' and Labourers' Dwellings—Public Conveniences—Scavenging, including Street Cleansing—Watering and the Removing of Snow—Planting Street Trees—Deposit of Plans—Dangerous Buildings—Hoardings—Obstructions—Improving Street Lines—Cellar Openings—Public Pleasure Grounds—Cemeteries—Mortuaries—Cattle and Ordinary Markets—Public Slaughter-houses, etc.—Giving numerous Forms of Notices, Specifications, and General Information upon these and other subjects of great importance to Municipal Engineers and others engaged in Sanitary Work.

Metrical Tables. By Sir G. L. MOLESWORTH, M.I.C.E. 32mo, cloth, 1s. 6d.

CONTENTS.

General—Linear Measures—Square Measures—Cubic Measures—Measures of Capacity—Weights—Combinations—Thermometers.

Elements of Construction for Electro-Magnets. By Count TH. DU MONCEL, Mem. de l'Institut de France. Translated from the French by C. J. WHARTON. Crown 8vo, cloth, 4s. 6d.

A Treatise on the Use of Belting for the Transmission of Power. By J. H. COOPER. Second edition, *illustrated,* 8vo, cloth, 15s.

A Pocket-Book of Useful Formulæ and Memoranda for Civil and Mechanical Engineers. By Sir GUILFORD L. MOLESWORTH, Mem. Inst. C.E. *With numerous illustrations,* 744 pp. Twenty-second edition, 32mo, roan, 6s.

SYNOPSIS OF CONTENTS:

Surveying, Levelling, etc.—Strength and Weight of Materials—Earthwork, Brickwork, Masonry, Arches, etc.—Struts, Columns, Beams, and Trusses—Flooring, Roofing, and Roof Trusses—Girders, Bridges, etc.—Railways and Roads—Hydraulic Formulæ—Canals, Sewers, Waterworks, Docks—Irrigation and Breakwaters—Gas, Ventilation, and Warming—Heat, Light, Colour, and Sound—Gravity: Centres, Forces, and Powers—Millwork, Teeth of Wheels, Shafting, etc.—Workshop Recipes—Sundry Machinery—Animal Power—Steam and the Steam Engine—Water-power, Water-wheels, Turbines, etc.—Wind and Windmills—Steam Navigation, Ship Building, Tonnage, etc.—Gunnery, Projectiles, etc.—Weights, Measures, and Money—Trigonometry, Conic Sections, and Curves—Telegraphy—Mensuration—Tables of Areas and Circumference, and Arcs of Circles—Logarithms, Square and Cube Roots, Powers—Reciprocals, etc.—Useful Numbers—Differential and Integral Calculus—Algebraic Signs—Telegraphic Construction and Formulæ.

Hints on Architectural Draughtsmanship. By G. W. TUXFORD HALLATT. Fcap. 8vo, cloth, 1s. 6d.

Spons' Tables and Memoranda for Engineers; selected and arranged by J. T. HURST, C.E., Author of 'Architectural Surveyors' Handbook,' 'Hurst's Tredgold's Carpentry,' etc. Eleventh edition, 64mo, roan, gilt edges, 1s.; or in cloth case, 1s. 6d.

This work is printed in a pearl type, and is so small, measuring only 2½ in. by 1¾ in. by ¼ in. thick, that it may be easily carried in the waistcoat pocket.

"It is certainly an extremely rare thing for a reviewer to be called upon to notice a volume measuring but 2½ in. by 1⅝ in., yet these dimensions faithfully represent the size of the handy little book before us. The volume—which contains 218 printed pages, besides a few blank pages for memoranda—is, in fact, a true pocket-book, adapted for being carried in the waistcoat pocket, and containing a far greater amount and variety of information than most people would imagine could be compressed into so small a space. . . . . The little volume has been compiled with considerable care and judgment, and we can cordially recommend it to our readers as a useful little pocket companion."—*Engineering.*

A Practical Treatise on Natural and Artificial Concrete, its Varieties and Constructive Adaptations. By HENRY REID, Author of the 'Science and Art of the Manufacture of Portland Cement.' New Edition, with 59 *woodcuts and* 5 *plates,* 8vo, cloth, 15s.

Notes on Concrete and Works in Concrete; especially written to assist those engaged upon Public Works. By JOHN NEWMAN, Assoc. Mem. Inst. C.E., crown 8vo, cloth, 4s. 6d.

Electricity as a Motive Power. By Count TH. DU MONCEL, Membre de l'Institut de France, and FRANK GERALDY, Ingénieur des Ponts et Chaussées. Translated and Edited, with Additions, by C. J. WHARTON, Assoc. Soc. Tel. Eng. and Elec. With 113 *engravings and diagrams,* crown 8vo, cloth, 7s. 6d.

Treatise on Valve-Gears, with special consideration of the Link-Motions of Locomotive Engines. By Dr. GUSTAV ZEUNER, Professor of Applied Mechanics at the Confederated Polytechnikum of Zurich. Translated from the Fourth German Edition, by Professor J. F. KLEIN, Lehigh University, Bethlehem, Pa. *Illustrated,* 8vo, cloth, 12s. 6d.

The French-Polisher's Manual. By a French-Polisher; containing Timber Staining, Washing, Matching, Improving, Painting, Imitations, Directions for Staining, Sizing, Embodying, Smoothing, Spirit Varnishing, French-Polishing, Directions for Re-polishing. Third edition, royal 32mo, sewed, 6d.

Hops, their Cultivation, Commerce, and Uses in various Countries. By P. L. SIMMONDS. Crown 8vo, cloth, 4s. 6d.

The Principles of Graphic Statics. By GEORGE SYDENHAM CLARKE, Major Royal Engineers. *With* 112 *illustrations.* Second edition, 4to, cloth, 12s. 6d.

Dynamo Tenders' Hand-Book. By F. B. BADT, late 1st Lieut. Royal Prussian Artillery. *With* 70 *illustrations.* Third edition, 18mo, cloth, 4s. 6d.

Practical Geometry, Perspective, and Engineering Drawing; a Course of Descriptive Geometry adapted to the Requirements of the Engineering Draughtsman, including the determination of cast shadows and Isometric Projection, each chapter being followed by numerous examples; to which are added rules for Shading, Shade-lining, etc., together with practical instructions as to the Lining, Colouring, Printing, and general treatment of Engineering Drawings, with a chapter on drawing Instruments. By GEORGE S. CLARKE, Capt. R.E. Second edition, *with* 21 *plates.* 2 vols., cloth, 10s. 6d.

The Elements of Graphic Statics. By Professor KARL VON OTT, translated from the German by G. S. CLARKE, Capt. R.E., Instructor in Mechanical Drawing, Royal Indian Engineering College. *With* 93 *illustrations*, crown 8vo, cloth, 5s.

A Practical Treatise on the Manufacture and Distribution of Coal Gas. By WILLIAM RICHARDS. Demy 4to, with *numerous wood engravings and* 29 *plates*, cloth, 28s.

SYNOPSIS OF CONTENTS:

Introduction—History of Gas Lighting—Chemistry of Gas Manufacture, by Lewis Thompson, Esq., M.R.C.S.—Coal, with Analyses, by J. Paterson, Lewis Thompson, and G. R. Hislop, Esqrs.—Retorts, Iron and Clay—Retort Setting—Hydraulic Main—Condensers—Exhausters—Washers and Scrubbers—Purifiers—Purification—History of Gas Holder—Tanks, Brick and Stone, Composite, Concrete, Cast-iron, Compound Annular Wrought-iron—Specifications—Gas Holders—Station Meter—Governor—Distribution—Mains—Gas Mathematics, or Formulæ for the Distribution of Gas, by Lewis Thompson, Esq.—Services—Consumers' Meters—Regulators—Burners—Fittings—Photometer—Carburization of Gas—Air Gas and Water Gas—Composition of Coal Gas, by Lewis Thompson, Esq.—Analyses of Gas—Influence of Atmospheric Pressure and Temperature on Gas—Residual Products—Appendix—Description of Retort Settings, Buildings, etc., etc.

The New Formula for Mean Velocity of Discharge of Rivers and Canals. By W. R. KUTTER. Translated from articles in the 'Cultur-Ingénieur,' by LOWIS D'A. JACKSON, Assoc. Inst. C.E. 8vo, cloth, 12s. 6d.

The Practical Millwright and Engineer's Ready Reckoner; or Tables for finding the diameter and power of cog-wheels, diameter, weight, and power of shafts, diameter and strength of bolts, etc. By THOMAS DIXON. Fourth edition, 12mo, cloth, 3s.

Tin: Describing the Chief Methods of Mining, Dressing and Smelting it abroad; with Notes upon Arsenic, Bismuth and Wolfram. By ARTHUR G. CHARLETON, Mem. American Inst. of Mining Engineers. *With plates*, 8vo, cloth, 12s. 6d.

B 3

Perspective, Explained and Illustrated. By G. S. CLARKE, Capt. R.E. *With illustrations*, 8vo, cloth, 3s. 6d.

Practical Hydraulics; a Series of Rules and Tables for the use of Engineers, etc., etc. By THOMAS BOX. Ninth edition, *numerous plates*, post 8vo, cloth, 5s.

The Essential Elements of Practical Mechanics; based on the Principle of Work, designed for Engineering Students. By OLIVER BYRNE, formerly Professor of Mathematics, College for Civil Engineers. Third edition, *with* 148 *wood engravings*, post 8vo, cloth, 7s. 6d.

CONTENTS:

Chap. 1. How Work is Measured by a Unit, both with and without reference to a Unit of Time—Chap. 2. The Work of Living Agents, the Influence of Friction, and introduces one of the most beautiful Laws of Motion—Chap. 3. The principles expounded in the first and second chapters are applied to the Motion of Bodies—Chap. 4. The Transmission of Work by simple Machines—Chap. 5. Useful Propositions and Rules.

Breweries and Maltings: their Arrangement, Construction, Machinery, and Plant. By G. SCAMELL, F.R.I.B.A. Second edition, revised, enlarged, and partly rewritten. By F. COLYER, M.I.C.E., M.I.M.E. *With* 20 *plates*, 8vo, cloth, 12s. 6d.

A Practical Treatise on the Construction of Horizontal and Vertical Waterwheels, specially designed for the use of operative mechanics. By WILLIAM CULLEN, Millwright and Engineer. *With* 11 *plates*. Second edition, revised and enlarged, small 4to, cloth, 12s. 6d.

A Practical Treatise on Mill-gearing, Wheels, Shafts, Riggers, etc.; for the use of Engineers. By THOMAS BOX. Third edition, *with* 11 *plates*. Crown 8vo, cloth, 7s. 6d.

Mining Machinery: a Descriptive Treatise on the Machinery, Tools, and other Appliances used in Mining. By G. G. ANDRÉ, F.G.S., Assoc. Inst. C.E., Mem. of the Society of Engineers. Royal 4to, uniform with the Author's Treatise on Coal Mining, containing 182 *plates*, accurately drawn to scale, with descriptive text, in 2 vols., cloth, 3l. 12s.

CONTENTS:

Machinery for Prospecting, Excavating, Hauling, and Hoisting—Ventilation—Pumping—Treatment of Mineral Products, including Gold and Silver, Copper, Tin, and Lead, Iron Coal, Sulphur, China Clay, Brick Earth, etc.

Tables for Setting out Curves for Railways, Canals, Roads, etc., varying from a radius of five chains to three miles. By A. KENNEDY and R. W. HACKWOOD. *Illustrated* 32mo, cloth, 2s. 6d.

Practical Electrical Notes and Definitions for the
use of Engineering Students and Practical Men. By W. PERREN
MAYCOCK, Assoc. M. Inst. E.E., Instructor in Electrical Engineering at
the Pitlake Institute, Croydon, together with the Rules and Regulations
to be observed in Electrical Installation Work. Royal 32mo, cloth.

The Draughtsman's Handbook of Plan and Map
Drawing; including instructions for the preparation of Engineering,
Architectural, and Mechanical Drawings. *With numerous illustrations
in the text, and* 33 *plates* (15 *printed in colours*). By G. G. ANDRÉ,
F.G.S., Assoc. Inst. C.E. 4to, cloth, 9s.

CONTENTS:
The Drawing Office and its Furnishings—Geometrical Problems—Lines, Dots, and their
Combinations—Colours, Shading, Lettering, Bordering, and North Points—Scales—Plotting
—Civil Engineers' and Surveyors' Plans—Map Drawing—Mechanical and Architectural
Drawing—Copying and Reducing Trigonometrical Formulæ, etc., etc.

The Boiler-maker's and Iron Ship-builder's Companion,
comprising a series of original and carefully calculated tables, of the
utmost utility to persons interested in the iron trades. By JAMES FODEN,
author of 'Mechanical Tables,' etc. Second edition revised, *with illustrations,* crown 8vo, cloth, 5s.

Rock Blasting: a Practical Treatise on the means
employed in Blasting Rocks for Industrial Purposes. By G. G. ANDRÉ,
F.G.S., Assoc. Inst. C.E. With 56 *illustrations and* 12 *plates,* 8vo, cloth,
10s. 6d.

Experimental Science: Elementary, Practical, and
Experimental Physics. By GEO. M. HOPKINS. *Illustrated by* 672
engravings. In one large vol., 8vo, cloth, 18s.

A Treatise on Ropemaking as practised in public and
private Rope-yards, with a Description of the Manufacture, Rules, Tables
of Weights, etc., adapted to the Trade, Shipping, Mining, Railways,
Builders, etc. By R. CHAPMAN, formerly foreman to Messrs. Huddart
and Co., Limehouse, and late Master Ropemaker to H.M. Dockyard,
Deptford. Second edition, 12mo, cloth, 3s.

Laxton's Builders' and Contractors' Tables; for the
use of Engineers, Architects, Surveyors, Builders, Land Agents, and
others. Bricklayer, containing 22 tables, with nearly 30,000 calculations.
4to, cloth, 5s.

Laxton's Builders' and Contractors' Tables. Excavator, Earth, Land, Water, and Gas, containing 53 tables, with nearly
24,000 calculations. 4to, cloth, 5s.

Egyptian Irrigation. By W. WILLCOCKS, M.I.C.E., Indian Public Works Department, Inspector of Irrigation, Egypt. With Introduction by Lieut.-Col. J. C. ROSS, R.E., Inspector-General of Irrigation. *With numerous lithographs and wood engravings*, royal 8vo, cloth, 1*l.* 16*s.*

Screw Cutting Tables for Engineers and Machinists, giving the values of the different trains of Wheels required to produce Screws of any pitch, calculated by Lord Lindsay, M.P., F.R.S., F.R.A.S., etc. Cloth, oblong, 2*s.*

Screw Cutting Tables, for the use of Mechanical Engineers, showing the proper arrangement of Wheels for cutting the Threads of Screws of any required pitch, with a Table for making the Universal Gas-pipe Threads and Taps. By W. A. MARTIN, Engineer. Second edition, oblong, cloth, 1*s.*, or sewed, 6*d.*

A Treatise on a Practical Method of Designing Slide-Valve Gears by Simple Geometrical Construction, based upon the principles enunciated in Euclid's Elements, and comprising the various forms of Plain Slide-Valve and Expansion Gearing; together with Stephenson's, Gooch's, and Allan's Link-Motions, as applied either to reversing or to variable expansion combinations. By EDWARD J. COWLING WELCH, Memb. Inst. Mechanical Engineers. Crown 8vo, cloth, 6*s.*

Cleaning and Scouring: a Manual for Dyers, Laundresses, and for Domestic Use. By S. CHRISTOPHER. 18mo, sewed, 6*d.*

A Glossary of Terms used in Coal Mining. By WILLIAM STUKELEY GRESLEY, Assoc. Mem. Inst. C.E., F.G.S., Member of the North of England Institute of Mining Engineers. *Illustrated with numerous woodcuts and diagrams*, crown 8vo, cloth, 5*s.*

A Pocket-Book for Boiler Makers and Steam Users, comprising a variety of useful information for Employer and Workman, Government Inspectors, Board of Trade Surveyors, Engineers in charge of Works and Slips, Foremen of Manufactories, and the general Steam-using Public. By MAURICE JOHN SEXTON. Second edition, royal 32mo, roan, gilt edges, 5*s.*

Electrolysis: a Practical Treatise on Nickeling, Coppering, Gilding, Silvering, the Refining of Metals, and the treatment of Ores by means of Electricity. By HIPPOLYTE FONTAINE, translated from the French by J. A. BERLY, C.E., Assoc. S.T.E. *With engravings.* 8vo, cloth, 9*s.*

Barlow's Tables of Squares, Cubes, Square Roots,
Cube Roots, Reciprocals of all Integer Numbers up to 10,000. Post 8vo, cloth, 6s.

A Practical Treatise on the Steam Engine, containing Plans and Arrangements of Details for Fixed Steam Engines, with Essays on the Principles involved in Design and Construction. By ARTHUR RIGG, Engineer, Member of the Society of Engineers and of the Royal Institution of Great Britain. Demy 4to, *copiously illustrated with woodcuts and 96 plates*, in one Volume, half-bound morocco, 2l. 2s.; or cheaper edition, cloth, 25s.

This work is not, in any sense, an elementary treatise, or history of the steam engine, but is intended to describe examples of Fixed Steam Engines without entering into the wide domain of locomotive or marine practice. To this end illustrations will be given of the most recent arrangements of Horizontal, Vertical, Beam, Pumping, Winding, Portable, Semi-portable, Corliss, Allen, Compound, and other similar Engines, by the most eminent Firms in Great Britain and America. The laws relating to the action and precautions to be observed in the construction of the various details, such as Cylinders, Pistons, Piston-rods, Connecting-rods, Cross-heads, Motion-blocks, Eccentrics, Simple, Expansion, Balanced, and Equilibrium Slide-valves, and Valve-gearing will be minutely dealt with. In this connection will be found articles upon the Velocity of Reciprocating Parts and the Mode of Applying the Indicator, Heat and Expansion of Steam Governors, and the like. It is the writer's desire to draw illustrations from every possible source, and give only those rules that present practice deems correct.

A Practical Treatise on the Science of Land and Engineering Surveying, Levelling, Estimating Quantities, etc., with a general description of the several Instruments required for Surveying, Levelling, Plotting, etc. By H. S. MERRETT. Fourth edition, revised by G. W. USILL, Assoc. Mem. Inst. C.E. 41 *plates, with illustrations and tables*, royal 8vo, cloth, 12s. 6d.

PRINCIPAL CONTENTS:

Part 1. Introduction and the Principles of Geometry. Part 2. Land Surveying; comprising General Observations—The Chain—Offsets Surveying by the Chain only—Surveying Hilly Ground—To Survey an Estate or Parish by the Chain only—Surveying with the Theodolite—Mining and Town Surveying—Railroad Surveying—Mapping—Division and Laying out of Land—Observations on Enclosures—Plane Trigonometry. Part 3. Levelling—Simple and Compound Levelling—The Level Book—Parliamentary Plan and Section—Levelling with a Theodolite—Gradients—Wooden Curves—To Lay out a Railway Curve—Setting out Widths. Part 4. Calculating Quantities generally for Estimates—Cuttings and Embankments—Tunnels—Brickwork—Ironwork—Timber Measuring. Part 5. Description and Use of Instruments in Surveying and Plotting—The Improved Dumpy Level—Troughton's Level—The Prismatic Compass—Proportional Compass—Box Sextant—Vernier—Pantagraph—Merrett's Improved Quadrant—Improved Computation Scale—The Diagonal Scale—Straight Edge and Sector. Part 6. Logarithms of Numbers—Logarithmic Sines and Co-Sines, Tangents and Co-Tangents—Natural Sines and Co-Sines—Tables for Earthwork, for Setting out Curves, and for various Calculations, etc., etc., etc.

Mechanical Graphics. A Second Course of Mechanical Drawing. With Preface by Prof. PERRY, B.Sc., F.R.S. Arranged for use in Technical and Science and Art Institutes, Schools and Colleges, by GEORGE HALLIDAY, Whitworth Scholar. 8vo, cloth, 6s.

B 4

The Assayer's Manual: an Abridged Treatise on the Docimastic Examination of Ores and Furnace and other Artificial Products. By BRUNO KERL. Translated by W. T. BRANNT. *With* 65 *illustrations*, 8vo, cloth, 12s. 6d.

Dynamo-Electric Machinery: a Text-Book for Students of Electro-Technology. By SILVANUS P. THOMPSON, B.A., D.Sc., M.S.T.E. [*New edition in the press.*]

The Practice of Hand Turning in Wood, Ivory, Shell, etc., with Instructions for Turning such Work in Metal as may be required in the Practice of Turning in Wood, Ivory, etc.; also an Appendix on Ornamental Turning. (A book for beginners.) By FRANCIS CAMPIN. Third edition, *with wood engravings*, crown 8vo, cloth, 6s.

CONTENTS:

On Lathes—Turning Tools—Turning Wood—Drilling—Screw Cutting—Miscellaneous Apparatus and Processes—Turning Particular Forms—Staining—Polishing—Spinning Metals—Materials—Ornamental Turning, etc.

Treatise on Watchwork, Past and Present. By the Rev. H. L. NELTHROPP, M.A., F.S.A. *With* 32 *illustrations*, crown 8vo, cloth, 6s. 6d.

CONTENTS:

Definitions of Words and Terms used in Watchwork—Tools—Time—Historical Summary—On Calculations of the Numbers for Wheels and Pinions; their Proportional Sizes, Trains, etc.—Of Dial Wheels, or Motion Work—Length of Time of Going without Winding up—The Verge—The Horizontal—The Duplex—The Lever—The Chronometer—Repeating Watches—Keyless Watches—The Pendulum, or Spiral Spring—Compensation—Jewelling of Pivot Holes—Clerkenwell—Fallacies of the Trade—Incapacity of Workmen—How to Choose and Use a Watch, etc.

Algebra Self-Taught. By W. P. HIGGS, M.A., D.Sc., LL.D., Assoc. Inst. C.E., Author of 'A Handbook of the Differential Calculus,' etc. Second edition, crown 8vo, cloth, 2s. 6d.

CONTENTS:

Symbols and the Signs of Operation—The Equation and the Unknown Quantity—Positive and Negative Quantities—Multiplication—Involution—Exponents—Negative Exponents—Roots, and the Use of Exponents as Logarithms—Logarithms—Tables of Logarithms and Proportionate Parts—Transformation of System of Logarithms—Common Uses of Common Logarithms—Compound Multiplication and the Binomial Theorem—Division, Fractions, and Ratio—Continued Proportion—The Series and the Summation of the Series—Limit of Series—Square and Cube Roots—Equations—List of Formulæ, etc.

Spons' Dictionary of Engineering, Civil, Mechanical, Military, and Naval; with technical terms in French, German, Italian, and Spanish, 3100 pp., and *nearly* 8000 *engravings*, in super-royal 8vo, in 8 divisions, 5*l*. 8*s*. Complete in 3 vols., cloth, 5*l*. 5*s*. Bound in a superior manner, half-morocco, top edge gilt, 3 vols., 6*l*. 12*s*.

Notes in Mechanical Engineering. Compiled principally for the use of the Students attending the Classes on this subject at the City of London College. By HENRY ADAMS, Mem. Inst. M.E. Mem. Inst. C.E., Mem. Soc. of Engineers. Crown 8vo, cloth, 2s. 6d.

Canoe and Boat Building: a complete Manual for Amateurs, containing plain and comprehensive directions for the construction of Canoes, Rowing and Sailing Boats, and Hunting Craft. By W. P. STEPHENS. *With numerous illustrations and 24 plates of Working Drawings.* Crown 8vo, cloth, 9s.

Proceedings of the National Conference of Electricians, *Philadelphia,* October 8th to 13th, 1884. 18mo, cloth, 3s.

Dynamo-Electricity, its Generation, Application, Transmission, Storage, and Measurement. By G. B. PRESCOTT. *With 545 illustrations.* 8vo, cloth, 1l. 1s.

Domestic Electricity for Amateurs. Translated from the French of E. HOSPITALIER, Editor of "L'Electricien," by C. J. WHARTON, Assoc. Soc. Tel. Eng. *Numerous illustrations.* Demy 8vo, cloth, 6s.

CONTENTS:

1. Production of the Electric Current—2. Electric Bells—3. Automatic Alarms—4. Domestic Telephones—5. Electric Clocks—6. Electric Lighters—7. Domestic Electric Lighting—8. Domestic Application of the Electric Light—9. Electric Motors—10. Electrical Locomotion—11. Electrotyping, Plating, and Gilding—12. Electric Recreations—13. Various applications—Workshop of the Electrician.

Wrinkles in Electric Lighting. By VINCENT STEPHEN. *With illustrations.* 18mo, cloth, 2s. 6d.

CONTENTS:

1. The Electric Current and its production by Chemical means—2. Production of Electric Currents by Mechanical means—3. Dynamo-Electric Machines—4. Electric Lamps—5. Lead—6. Ship Lighting.

Foundations and Foundation Walls for all classes of Buildings, Pile Driving, Building Stones and Bricks, Pier and Wall construction, Mortars, Limes, Cements, Concretes, Stuccos, &c. 64 *illustrations.* By G. T. POWELL and F. BAUMAN. 8vo, cloth, 10s. 6d.

Manual for Gas Engineering Students. By D. LEE. 18mo, cloth, 1s.

Hydraulic Machinery, Past and Present. A Lecture delivered to the London and Suburban Railway Officials' Association. By H. ADAMS, Mem. Inst. C.E. *Folding plate.* 8vo, sewed, 1s.

Twenty Years with the Indicator. By THOMAS PRAY, Jun., C.E., M.E., Member of the American Society of Civil Engineers. 2 vols., royal 8vo, cloth, 12s. 6d.

Annual Statistical Report of the Secretary to the Members of the Iron and Steel Association on the Home and Foreign Iron and Steel Industries in 1889. Issued June 1890. 8vo, sewed, 5s.

Bad Drains, and How to Test them; with Notes on the Ventilation of Sewers, Drains, and Sanitary Fittings, and the Origin and Transmission of Zymotic Disease. By R. HARRIS REEVES. Crown 8vo, cloth, 3s. 6d.

Well Sinking. The modern practice of Sinking and Boring Wells, with geological considerations and examples of Wells. By ERNEST SPON, Assoc. Mem. Inst. C.E., Mem. Soc. Eng., and of the Franklin Inst., etc. Second edition, revised and enlarged. Crown 8vo, cloth, 10s. 6d.

The Voltaic Accumulator: an Elementary Treatise. By ÉMILE REYNIER. Translated by J. A. BERLY, Assoc. Inst. E.E. *With 62 illustrations,* 8vo, cloth, 9s.

List of Tests (Reagents), arranged in alphabetical order, according to the names of the originators. Designed especially for the convenient reference of Chemists, Pharmacists, and Scientists. By HANS M. WILDER. Crown 8vo, cloth, 4s. 6d.

Ten Years' Experience in Works of Intermittent Downward Filtration. By J. BAILEY DENTON, Mem. Inst. C.E. Second edition, with additions. Royal 8vo, sewed, 4s.

A Treatise on the Manufacture of Soap and Candles, Lubricants and Glycerin. By W. LANT CARPENTER, B.A., B.Sc. (late of Messrs. C. Thomas and Brothers, Bristol). *With illustrations.* Crown 8vo, cloth, 10s. 6d.

Land Surveying on the Meridian and Perpendicular System. By WILLIAM PENMAN, C.E. 8vo, cloth, 8s. 6d.

Incandescent Wiring Hand-Book. By F. B. BADT, late 1st Lieut. Royal Prussian Artillery. *With* 41 *illustrations and* 5 *tables.* 18mo, cloth, 4s. 6d.

*A Pocket-book for Pharmacists, Medical Prac-*titioners, Students, *etc.,* etc. (*British, Colonial, and American*). By THOMAS BAYLEY, Assoc. R. Coll. of Science, Consulting Chemist, Analyst, and Assayer, Author of a 'Pocket-book for Chemists,' 'The Assay and Analysis of Iron and Steel, Iron Ores, and Fuel,' etc., etc. Royal 32mo, boards, gilt edges, 6s.

The Fireman's Guide; a Handbook on the Care of Boilers. By TEKNOLOG, föreningen T. I. Stockholm. Translated from the third edition, and revised by KARL P. DAHLSTROM, M.E. Second edition. Fcap. 8vo, cloth, 2s.

A Treatise on Modern Steam Engines and Boilers, including Land Locomotive, and Marine Engines and Boilers, for the use of Students. By FREDERICK COLYER, M. Inst. C.E., Mem. Inst. M.E. *With* 36 *plates.* 4to, cloth, 12s. 6d.

CONTENTS:

1. Introduction—2. Original Engines—3. Boilers—4. High-Pressure Beam Engines—5. Cornish Beam Engines—6. Horizontal Engines—7. Oscillating Engines—8. Vertical High-Pressure Engines—9. Special Engines—10. Portable Engines—11. Locomotive Engines—12. Marine Engines.

Steam Engine Management; a Treatise on the Working and Management of Steam Boilers. By F. COLYER, M. Inst. C.E., Mem. Inst. M.E. 18mo, cloth, 2s.

A Text-Book of Tanning, embracing the Preparation of all kinds of Leather. By HARRY R. PROCTOR, F.C.S., of Low Lights Tanneries. *With illustrations.* Crown 8vo, cloth, 10s. 6d.

Aid Book to Engineering Enterprise. By EWING MATHESON, M. Inst. C.E. The Inception of Public Works, Parliamentary Procedure for Railways, Concessions for Foreign Works, and means of Providing Money, the Points which determine Success or Failure, Contract and Purchase, Commerce in Coal, Iron, and Steel, &c. Second edition, revised and enlarged, 8vo, cloth, 21s.

Pumps, Historically, Theoretically, and Practically Considered. By P. R. BJÖRLING. *With 156 illustrations.* Crown 8vo, cloth, 7s. 6d.

The Marine Transport of Petroleum. A Book for the use of Shipowners, Shipbuilders, Underwriters, Merchants, Captains and Officers of Petroleum-carrying Vessels. By G. H. LITTLE, Editor of the 'Liverpool Journal of Commerce.' Crown 8vo, cloth, 10s. 6d.

Liquid Fuel for Mechanical and Industrial Purposes. Compiled by E. A. BRAYLEY HODGETTS. *With wood engravings.* 8vo, cloth, 7s. 6d.

Tropical Agriculture: A Treatise on the Culture, Preparation, Commerce and Consumption of the principal Products of the Vegetable Kingdom. By P. L. SIMMONDS, F.L.S., F.R.C.I. New edition, revised and enlarged, 8vo, cloth, 21s.

Health and Comfort in House Building; or, Ventilation with Warm Air by Self-acting Suction Power. With Review of the Mode of Calculating the Draught in Hot-air Flues, and with some Actual Experiments by J. DRYSDALE, M.D., and J. W. HAYWARD, M.D. *With plates and woodcuts.* Third edition, with some New Sections, and the whole carefully Revised, 8vo, cloth, 7s. 6d.

Losses in Gold Amalgamation. With Notes on the Concentration of Gold and Silver Ores. *With six plates.* By W. McDERMOTT and P. W. DUFFIELD. 8vo, cloth, 5s.

A Guide for the Electric Testing of Telegraph Cables. By Col. V. HOSKIŒR, Royal Danish Engineers. Third edition, crown 8vo, cloth, 4s. 6d.

The Hydraulic Gold Miners' Manual. By T. S. G. KIRKPATRICK, M.A. Oxon. *With 6 plates.* Crown 8vo, cloth, 6s.

Irrigation Manual. By Lieut.-Gen. J. MULLINS, Royal (late Madras) Engineers, retired; sometime Chief Engineer for Irrigation, Madras, and Fellow of the University of Madras. *With numerous plates and tables.* Published for the Madras Government. Small folio, cloth or half-bound calf, 4l. 4s.

The Turkish Bath: Its Design and Construction for Public and Commercial Purposes. By R. O. ALLSOP, Architect. *With plans and sections.* 8vo, cloth, 6s.

Earthwork Slips and Subsidences upon Public Works: Their Causes, Prevention and Reparation. Especially written to assist those engaged in the Construction or Maintenance of Railways, Docks, Canals, Waterworks, River Banks, Reclamation Embankments, Drainage Works, &c., &c. By JOHN NEWMAN, Assoc. Mem. Inst. C.E., Author of 'Notes on Concrete,' &c. Crown 8vo, cloth, 7s. 6d.

Gas and Petroleum Engines: A Practical Treatise on the Internal Combustion Engine. By WM. ROBINSON, M.E., Senior Demonstrator and Lecturer on Applied Mechanics, Physics, &c., City and Guilds of London College, Finsbury, Assoc. Mem. Inst. C.E., &c. *Numerous illustrations.* 8vo, cloth, 14s.

Waterways and Water Transport in Different Countries. With a description of the Panama, Suez, Manchester, Nicaraguan, and other Canals. By J. STEPHEN JEANS, Author of 'England's Supremacy,' 'Railway Problems,' &c. *Numerous illustrations.* 8vo, cloth, 14s.

A Treatise on the Richards Steam-Engine Indicator and the Development and Application of Force in the Steam-Engine. By CHARLES T. PORTER. Fourth Edition, revised and enlarged, 8vo, cloth, 9s.

CONTENTS.

The Nature and Use of the Indicator: The several lines on the Diagram. Examination of Diagram No. 1.
Of Truth in the Diagram.
Description of the Richards Indicator.
Practical Directions for Applying and Taking Care of the Indicator.
Introductory Remarks.
Units.
Expansion.
Directions for ascertaining from the Diagram the Power exerted by the Engine.
To Measure from the Diagram the Quantity of Steam Consumed.
To Measure from the Diagram the Quantity of Heat Expended.
Of the Real Diagram, and how to Construct it.
Of the Conversion of Heat into Work in the Steam-engine.
Observations on the several Lines of the Diagram.
Of the Loss attending the Employment of Slow-piston Speed, and the Extent to which this is Shown by the Indicator.
Of other Applications of the Indicator.
Of the use of the Tables of the Properties of Steam in Calculating the Duty of Boilers. Introductory.
Of the Pressure on the Crank when the Connecting-rod is conceived to be of Infinite Length.
The Modification of the Acceleration and Retardation that is occasioned by the Angular Vibration of the Connecting-rod.
Method of representing the actual pressure on the crank at every point of its revolution.
The Rotative Effect of the Pressure exerted on the Crank.
The Transmitting Parts of an Engine, considered as an Equaliser of Motion.
A Ride on a Buffer-beam (Appendix).

In demy 4to, handsomely bound in cloth, *illustrated with* **220** *full page plates*, Price 15s.

ARCHITECTURAL EXAMPLES
IN BRICK, STONE, WOOD, AND IRON.
A COMPLETE WORK ON THE DETAILS AND ARRANGEMENT OF BUILDING CONSTRUCTION AND DESIGN.

By WILLIAM FULLERTON, Architect.

Containing 220 Plates, with numerous Drawings selected from the Architecture of Former and Present Times.

The Details and Designs are Drawn to Scale, $\frac{1}{8}''$, $\frac{1}{4}''$, $\frac{1}{2}''$, and Full size being chiefly used.

The Plates are arranged in Two Parts. The First Part contains Details of Work in the four principal Building materials, the following being a few of the subjects in this Part :—Various forms of Doors and Windows, Wood and Iron Roofs, Half Timber Work, Porches, Towers, Spires, Belfries, Flying Buttresses, Groining, Carving, Church Fittings, Constructive and Ornamental Iron Work, Classic and Gothic Molds and Ornament, Foliation Natural and Conventional, Stained Glass, Coloured Decoration, a Section to Scale of the Great Pyramid, Grecian and Roman Work, Continental and English Gothic, Pile Foundations, Chimney Shafts according to the regulations of the London County Council, Board Schools. The Second Part consists of Drawings of Plans and Elevations of Buildings, arranged under the following heads :—Workmen's Cottages and Dwellings, Cottage Residences and Dwelling Houses, Shops, Factories, Warehouses, Schools, Churches and Chapels, Public Buildings, Hotels and Taverns, and Buildings of a general character.

All the Plates are accompanied with particulars of the Work, with Explanatory Notes and Dimensions of the various parts.

Specimen Pages, reduced from the originals.

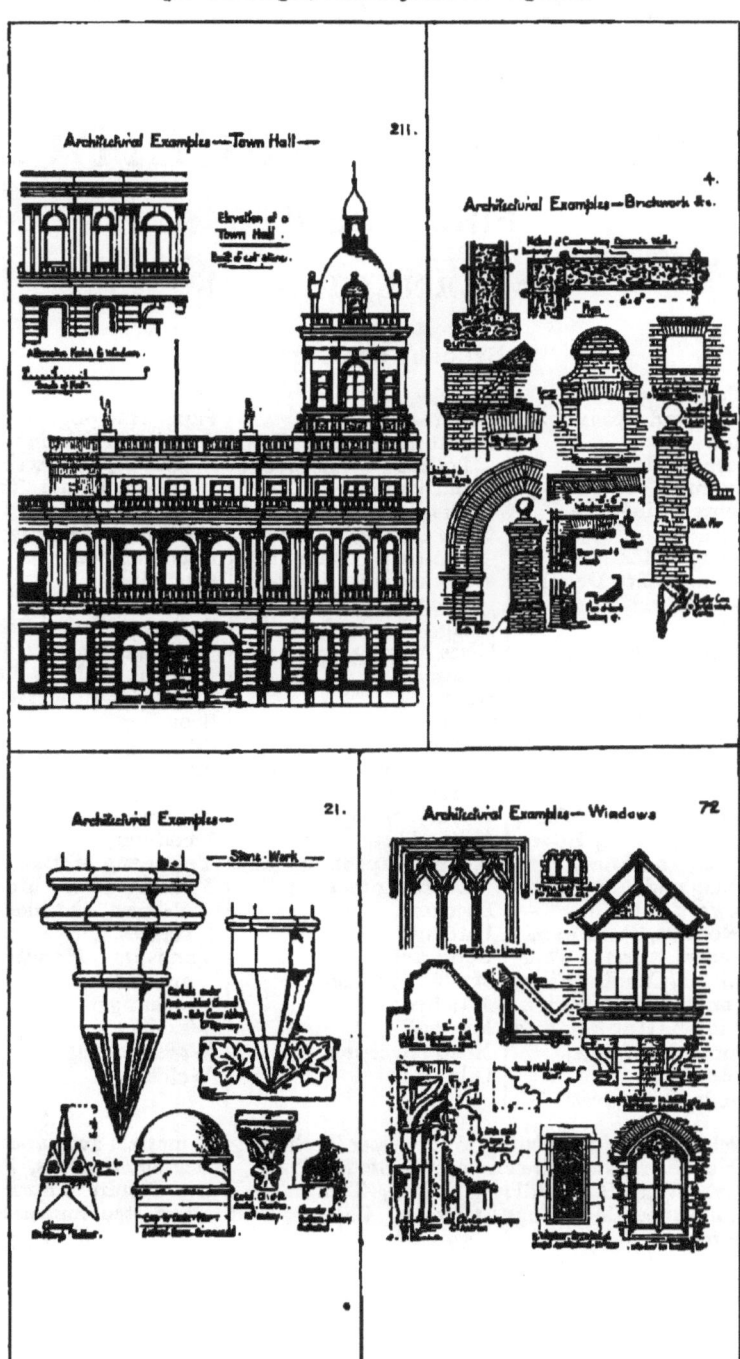

Crown 8vo, cloth, with illustrations, 5s.

WORKSHOP RECEIPTS,

FIRST SERIES.

By ERNEST SPON.

SYNOPSIS OF CONTENTS.

Bookbinding.
Bronzes and Bronzing.
Candles.
Cement.
Cleaning.
Colourwashing.
Concretes.
Dipping Acids.
Drawing Office Details.
Drying Oils.
Dynamite.
Electro - Metallurgy — (Cleaning, Dipping, Scratch-brushing, Batteries, Baths, and Deposits of every description).
Enamels.
Engraving on Wood, Copper, Gold, Silver, Steel, and Stone.
Etching and Aqua Tint.
Firework Making — (Rockets, Stars, Rains, Gerbes, Jets, Tourbillons, Candles, Fires, Lances, Lights, Wheels, Fire-balloons, and minor Fireworks).
Fluxes.
Foundry Mixtures.

Freezing.
Fulminates.
Furniture Creams, Oils, Polishes, Lacquers, and Pastes.
Gilding.
Glass Cutting, Cleaning, Frosting, Drilling, Darkening, Bending, Staining, and Painting.
Glass Making.
Glues.
Gold.
Graining.
Gums.
Gun Cotton.
Gunpowder.
Horn Working.
Indiarubber.
Japans, Japanning, and kindred processes.
Lacquers.
Lathing.
Lubricants.
Marble Working.
Matches.
Mortars.
Nitro-Glycerine.
Oils.

Paper.
Paper Hanging.
Painting in Oils, in Water Colours, as well as Fresco, House, Transparency, Sign, and Carriage Painting.
Photography.
Plastering.
Polishes.
Pottery—(Clays, Bodies, Glazes, Colours, Oils, Stains, Fluxes, Enamels, and Lustres).
Scouring.
Silvering.
Soap.
Solders.
Tanning.
Taxidermy.
Tempering Metals.
Treating Horn, Mother-o'-Pearl, and like substances.
Varnishes, Manufacture and Use of.
Veneering.
Washing.
Waterproofing.
Welding.

Besides Receipts relating to the lesser Technological matters and processes, such as the manufacture and use of Stencil Plates, Blacking, Crayons, Paste, Putty, Wax, Size, Alloys, Catgut, Tunbridge Ware, Picture Frame and Architectural Mouldings, Compos, Cameos, and others too numerous to mention.

Crown 8vo, cloth, 485 pages, with illustrations, 5s.

WORKSHOP RECEIPTS,
SECOND SERIES.
By ROBERT HALDANE.

Synopsis of Contents.

Acidimetry and Alkalimetry.	Disinfectants.	Iodoform.
Albumen.	Dyeing, Staining, and Colouring.	Isinglass.
Alcohol.	Essences.	Ivory substitutes.
Alkaloids.	Extracts.	Leather.
Baking-powders.	Fireproofing.	Luminous bodies.
Bitters.	Gelatine, Glue, and Size.	Magnesia.
Bleaching.	Glycerine.	Matches.
Boiler Incrustations.	Gut.	Paper.
Cements and Lutes.	Hydrogen peroxide.	Parchment.
Cleansing.	Ink.	Perchloric acid.
Confectionery.	Iodine.	Potassium oxalate.
Copying.		Preserving.

Pigments, Paint, and Painting: embracing the preparation of *Pigments*, including alumina lakes, blacks (animal, bone, Frankfort, ivory, lamp, sight, soot), blues (antimony, Antwerp, cobalt, cæruleum, Egyptian, manganate, Paris, Péligot, Prussian, smalt, ultramarine), browns (bistre, hinau, sepia, sienna, umber, Vandyke), greens (baryta, Brighton, Brunswick, chrome, cobalt, Douglas, emerald, manganese, mitis, mountain, Prussian, sap, Scheele's, Schweinfurth, titanium, verdigris, zinc), reds (Brazilwood lake, carminated lake, carmine, Cassius purple, cobalt pink, cochineal lake, colcothar, Indian red, madder lake, red chalk, red lead, vermilion), whites (alum, baryta, Chinese, lead sulphate, white lead—by American, Dutch, French, German, Kremnitz, and Pattinson processes, precautions in making, and composition of commercial samples—whiting, Wilkinson's white, zinc white), yellows (chrome, gamboge, Naples, orpiment, realgar, yellow lakes); *Paint* (vehicles, testing oils, driers, grinding, storing, applying, priming, drying, filling, coats, brushes, surface, water-colours, removing smell, discoloration; miscellaneous paints—cement paint for carton-pierre, copper paint, gold paint, iron paint, lime paints, silicated paints, steatite paint, transparent paints, tungsten paints, window paint, zinc paints); *Painting* (general instructions, proportions of ingredients, measuring paint work; carriage painting—priming paint, best putty, finishing colour, cause of cracking, mixing the paints, oils, driers, and colours, varnishing, importance of washing vehicles, re-varnishing, how to dry paint; woodwork painting).

Crown 8vo, cloth, 480 pages, with 183 illustrations, 5s.

WORKSHOP RECEIPTS,

THIRD SERIES.

By C. G. WARNFORD LOCK.

Uniform with the First and Second Series.

SYNOPSIS OF CONTENTS.

Alloys.	Indium.	Rubidium.
Aluminium.	Iridium.	Ruthenium.
Antimony.	Iron and Steel.	Selenium.
Barium.	Lacquers and Lacquering.	Silver.
Beryllium.	Lanthanum.	Slag.
Bismuth.	Lead.	Sodium.
Cadmium.	Lithium.	Strontium.
Cæsium.	Lubricants.	Tantalum.
Calcium.	Magnesium.	Terbium.
Cerium.	Manganese.	Thallium.
Chromium.	Mercury.	Thorium.
Cobalt.	Mica.	Tin.
Copper.	Molybdenum.	Titanium.
Didymium.	Nickel.	Tungsten.
Electrics.	Niobium.	Uranium.
Enamels and Glazes.	Osmium.	Vanadium.
Erbium.	Palladium.	Yttrium.
Gallium.	Platinum.	Zinc.
Glass.	Potassium.	Zirconium.
Gold.	Rhodium.	

PUBLISHED BY E. & F. N. SPON.

WORKSHOP RECEIPTS,
FOURTH SERIES,
DEVOTED MAINLY TO HANDICRAFTS & MECHANICAL SUBJECTS.

By C. G. WARNFORD LOCK.

250 Illustrations, with Complete Index, and a General Index to the Four Series, 5s.

Waterproofing — rubber goods, cuprammonium processes, miscellaneous preparations.
Packing and Storing articles of delicate odour or colour, of a deliquescent character, liable to ignition, apt to suffer from insects or damp, or easily broken.
Embalming and Preserving anatomical specimens.
Leather Polishes.
Cooling Air and Water, producing low temperatures, making ice, cooling syrups and solutions, and separating salts from liquors by refrigeration.
Pumps and Siphons, embracing every useful contrivance for raising and supplying water on a moderate scale, and moving corrosive, tenacious, and other liquids.
Desiccating—air- and water-ovens, and other appliances for drying natural and artificial products.
Distilling—water, tinctures, extracts, pharmaceutical preparations, essences, perfumes, and alcoholic liquids.
Emulsifying as required by pharmacists and photographers.
Evaporating—saline and other solutions, and liquids demanding special precautions.
Filtering—water, and solutions of various kinds.
Percolating and Macerating.
Electrotyping.
Stereotyping by both plaster and paper processes.
Bookbinding in all its details.
Straw Plaiting and the fabrication of baskets, matting, etc.
Musical Instruments—the preservation, tuning, and repair of pianos, harmoniums, musical boxes, etc.
Clock and Watch Mending—adapted for intelligent amateurs.
Photography—recent development in rapid processes, handy apparatus, numerous recipes for sensitizing and developing solutions, and applications to modern illustrative purposes.

NOW COMPLETE.

With nearly 1500 *illustrations*, in super-royal 8vo, in 5 Divisions, cloth. Divisions 1 to 4, 13s. 6d. each; Division 5, 17s. 6d.; or 2 vols., cloth, £3 10s.

SPONS' ENCYCLOPÆDIA
OF THE
INDUSTRIAL ARTS, MANUFACTURES, AND COMMERCIAL PRODUCTS.

EDITED BY C. G. WARNFORD LOCK, F.L.S.

Among the more important of the subjects treated of, are the following:—

Acids, 207 pp. 220 figs.
Alcohol, 23 pp. 16 figs.
Alcoholic Liquors, 13 pp.
Alkalies, 89 pp. 78 figs.
Alloys. Alum.
Asphalt. Assaying.
Beverages, 89 pp. 29 figs.
Blacks.
Bleaching Powder, 15 pp.
Bleaching, 51 pp. 48 figs.
Candles, 18 pp. 9 figs.
Carbon Bisulphide.
Celluloid, 9 pp.
Cements. Clay.
Coal-tar Products, 44 pp. 14 figs.
Cocoa, 8 pp.
Coffee, 32 pp. 13 figs.
Cork, 8 pp. 17 figs.
Cotton Manufactures, 62 pp. 57 figs.
Drugs, 38 pp.
Dyeing and Calico Printing, 28 pp. 9 figs.
Dyestuffs, 16 pp.
Electro-Metallurgy, 13 pp.
Explosives, 22 pp. 33 figs.
Feathers.
Fibrous Substances, 92 pp. 79 figs.
Floor-cloth, 16 pp. 21 figs.
Food Preservation, 8 pp.
Fruit, 8 pp.

Fur, 5 pp.
Gas, Coal, 8 pp.
Gems.
Glass, 45 pp. 77 figs.
Graphite, 7 pp.
Hair, 7 pp.
Hair Manufactures.
Hats, 26 pp. 26 figs.
Honey. Hops.
Horn.
Ice, 10 pp. 14 figs.
Indiarubber Manufactures, 23 pp. 17 figs.
Ink, 17 pp.
Ivory.
Jute Manufactures, 11 pp., 11 figs.
Knitted Fabrics — Hosiery, 15 pp. 13 figs.
Lace, 13 pp. 9 figs.
Leather, 28 pp. 31 figs.
Linen Manufactures, 16 pp. 6 figs.
Manures, 21 pp. 30 figs.
Matches, 17 pp. 38 figs.
Mordants, 13 pp.
Narcotics, 47 pp.
Nuts, 10 pp.
Oils and Fatty Substances, 125 pp.
Paint.
Paper, 26 pp. 23 figs.
Paraffin, 8 pp. 6 figs.
Pearl and Coral, 8 pp.
Perfumes, 10 pp.

Photography, 13 pp. 20 figs.
Pigments, 9 pp. 6 figs.
Pottery, 46 pp. 57 figs.
Printing and Engraving, 20 pp. 8 figs.
Rags.
Resinous and Gummy Substances, 75 pp. 16 figs.
Rope, 16 pp. 17 figs.
Salt, 31 pp. 23 figs.
Silk, 8 pp.
Silk Manufactures, 9 pp. 11 figs.
Skins, 5 pp.
Small Wares, 4 pp.
Soap and Glycerine, 39 pp. 45 figs.
Spices, 16 pp.
Sponge, 5 pp.
Starch, 9 pp. 10 figs.
Sugar, 155 pp. 134 figs.
Sulphur.
Tannin, 18 pp.
Tea, 12 pp.
Timber, 13 pp.
Varnish, 15 pp.
Vinegar, 5 pp.
Wax, 5 pp.
Wool, 2 pp.
Woollen Manufactures, 58 pp. 39 figs.

In super-royal 8vo, 1168 pp., *with* 2400 *illustrations*, in 3 Divisions, cloth, price 13*s.* 6*d.* each ; or 1 vol., cloth, 2*l.* ; or half-morocco, 2*l.* 8*s.*

A SUPPLEMENT

TO

SPONS' DICTIONARY OF ENGINEERING.

EDITED BY ERNEST SPON, MEMB. SOC. ENGINEERS.

Abacus, Counters, Speed Indicators, and Slide Rule.
Agricultural Implements and Machinery.
Air Compressors.
Animal Charcoal Machinery.
Antimony.
Axles and Axle-boxes.
Barn Machinery.
Belts and Belting.
Blasting. Boilers.
Brakes.
Brick Machinery.
Bridges.
Cages for Mines.
Calculus, Differential and Integral.
Canals.
Carpentry.
Cast Iron.
Cement, Concrete, Limes, and Mortar.
Chimney Shafts.
Coal Cleansing and Washing.

Coal Mining.
Coal Cutting Machines.
Coke Ovens. Copper.
Docks. Drainage.
Dredging Machinery.
Dynamo - Electric and Magneto-Electric Machines.
Dynamometers.
Electrical Engineering, Telegraphy, Electric Lighting and its practical details, Telephones
Engines, Varieties of.
Explosives. Fans.
Founding, Moulding and the practical work of the Foundry.
Gas, Manufacture of.
Hammers, Steam and other Power.
Heat. Horse Power.
Hydraulics.
Hydro-geology.
Indicators. Iron.
Lifts, Hoists, and Elevators.

Lighthouses, Buoys, and Beacons.
Machine Tools.
Materials of Construction.
Meters.
Ores, Machinery and Processes employed to Dress.
Piers.
Pile Driving.
Pneumatic Transmission.
Pumps.
Pyrometers.
Road Locomotives.
Rock Drills.
Rolling Stock.
Sanitary Engineering.
Shafting.
Steel.
Steam Navvy.
Stone Machinery.
Tramways.
Well Sinking.

JUST PUBLISHED.

In demy 8vo, cloth, 600 pages, and 1420 Illustrations, 6s.

SPONS'
MECHANICS' OWN BOOK;
A MANUAL FOR HANDICRAFTSMEN AND AMATEURS.

CONTENTS.

Mechanical Drawing—Casting and Founding in Iron, Brass, Bronze, and other Alloys—Forging and Finishing Iron—Sheetmetal Working—Soldering, Brazing, and Burning—Carpentry and Joinery, embracing descriptions of some 400 Woods, over 200 Illustrations of Tools and their uses, Explanations (with Diagrams) of 116 joints and hinges, and Details of Construction of Workshop appliances, rough furniture, Garden and Yard Erections, and House Building—Cabinet-Making and Veneering—Carving and Fretcutting—Upholstery—Painting, Graining, and Marbling—Staining Furniture, Woods, Floors, and Fittings—Gilding, dead and bright, on various grounds—Polishing Marble, Metals, and Wood—Varnishing—Mechanical movements, illustrating contrivances for transmitting motion—Turning in Wood and Metals—Masonry, embracing Stonework, Brickwork, Terracotta, and Concrete—Roofing with Thatch, Tiles, Slates, Felt, Zinc, &c.—Glazing with and without putty, and lead glazing—Plastering and Whitewashing—Paper-hanging—Gas-fitting—Bell-hanging, ordinary and electric Systems—Lighting—Warming—Ventilating—Roads, Pavements, and Bridges—Hedges, Ditches, and Drains—Water Supply and Sanitation—Hints on House Construction suited to new countries.

E. & F. N. SPON, 125, Strand, London.
New York: 12, Cortlandt Street.

www.ingramcontent.com/pod-product-compliance
Lightning Source LLC
Chambersburg PA
CBHW030303170426
43202CB00009B/857